POSTHUMANISM

POSTHUMANISM Alan Smart and Josephine Smart

ANTHROPOLOGICAL INSIGHTS

UNIVERSITY OF TORONTO PRESS

Copyright © University of Toronto Press 2017
Higher Education Division

www.utppublishing.com

All rights reserved. The use of any part of this publication reproduced, transmitted in any form or by any means, electronic, mechanical, photocopying, recording, or otherwise, or stored in a retrieval system, without prior written consent of the publisher—or in the case of photocopying, a licence from Access Copyright (the Canadian Copyright Licensing Agency), 320–56 Wellesley Street West, Toronto, Ontario, M5S 2S3—is an infringement of the copyright law.

Library and Archives Canada Cataloguing in Publication

Smart, Alan, 1956–, author
    Posthumanism / Alan Smart and Josephine Smart.

(Anthropological insights)
Includes bibliographical references and index.
Issued in print and electronic formats.

ISBN 978-1-4426-3642-2 (cloth).—ISBN 978-1-4426-3641-5 (paper).—
ISBN 978-1-4426-3644-6 (HTML).—ISBN 978-1-4426-3643-9 (PDF)

    1. Humanism.   2. Holism.   3. Anthropology.   4. Ethnology.
I. Smart, Josephine, author   II. Title.   III. Series: Anthropological insights

B821.S63 2017   149   C2016-906634-7
                                  C2016-906635-5

We welcome comments and suggestions regarding any aspect of our publications—please feel free to contact us at news@utphighereducation.com or visit our Internet site at www.utppublishing.com.

North America
5201 Dufferin Street
North York, Ontario, Canada, M3H 5T8

2250 Military Road
Tonawanda, New York, USA, 14150

ORDERS PHONE: 1-800-565-9523
ORDERS FAX: 1-800-221-9985
ORDERS E-MAIL: utpbooks@utpress.utoronto.ca

UK, Ireland, and continental Europe
NBN International
Estover Road, Plymouth, PL6 7PY, UK

ORDERS PHONE: 44 (0) 1752 202301
ORDERS FAX: 44 (0) 1752 202333
ORDERS E-MAIL:
enquiries@nbninternational.com

Every effort has been made to contact copyright holders; in the event of an error or omission, please notify the publisher.

The University of Toronto Press acknowledges the financial support for its publishing activities of the Government of Canada through the Canada Book Fund.

Printed in the United States of America.
Cover design: Grace Cheong.

This book is dedicated to our doctoral supervisors at the University of Toronto, Peter Carstens (1929–2010) and Shuichi Nagata (1931–2016)

CONTENTS

Acknowledgments ix
1 Posthumanism 1
2 Zoonotic Diseases and the Microbiome 17
3 Multispecies Ethnography 43
4 Technology, Cyborgs, and Transhumanism 65
Conclusion 95
Glossary 99
References 105
Index 115

ACKNOWLEDGMENTS

Although this book was mostly written in the second half of 2015, it has drawn on discussions and research carried out for over a decade. Our interest in posthumanism derived from research on the impact of bovine spongiform encephalopathy (BSE) funded by the Alberta Prion Research Institute and the Social Sciences and Humanities Research Council. Ideas related to this book were first tried out in a number of conference and workshop presentations. We benefited from the insightful suggestions of a number of anonymous reviewers of earlier drafts of this book and our related articles and book chapters, and several editors of books in which we included related chapters: Kendra Coulter, William R. Schumann, Thomas Wilson, Hastings Donnan, and Noel Salazar. Our engagement with these ideas was encouraged by our experience of being in a department where biological and social/cultural anthropologists cooperated and supported each other's endeavors; we express particular gratitude to the primatologists: Jim Paterson, Mary Pavelka, Pamela Asquith, Pascale Sicotte, Linda Fedigan, and Steig Johnson. There are too many people with whom we have discussed these issues to be able to mention them all, but we particularly want to thank Gwendolyn Blue, Susy Cote, Dean Curran, Agustin Fuentes, John Galaty, Paul Hansen, Tracey Heatherington, Josiah Heyman, Olivier LaRoque, Eliza Yee-ha Lok, Melanie Rock, Roy Runzer, Sheila Runzer, Eric Po-ming Yau, and Filippo Zerilli. The comments on draft chapters and discussion by the students in the posthumanism seminar in fall 2015—Dylan Archer, Heather Lee, Anh Ly, Katrina Nethercott, Katrina Palad, Jordan Renz, Taylor Shewchuk, Jade Wright, and Bertha Wu—were particularly helpful. Cara Tremain helped considerably with efficient editorial assistance. Finally, this project would not have been possible without the encouragement and useful advice of Anne Brackenbury and others at University of Toronto Press.

CHAPTER 1

POSTHUMANISM

Only about ten per cent of cells in the human body contain human DNA: most of the rest are part of a vast community of companion species, particularly bacteria and viruses, which recently has become the subject of intense medical interest and relabeled the **microbiome**. **Fecal transplants** are being experimented with as a way of treating a variety of illnesses by reintroducing important good bugs. Some scientists believe that autoimmune diseases such as Crohn's syndrome are the result of the control of parasitic worms, inducing the body to attack itself in the absence of its traditional enemies. But it is not only our bodies: some research suggests that adjusting the microbiome may be an effective technique to respond to depression and schizophrenia.

The US Department of Defense is engaged in research to design the soldier of the future. One current project is trying to create powered body armor "like Iron Man's" that would greatly enhance a soldier's strength and perceptual capacities. Another is developing an integrated platform that will provide organic and shared sensor information to soldiers. In the future, there is the potential for other enhancements through genetic or pharmaceutical modification, or the implantation of sensors and pumps for adrenaline and other drugs that could be triggered remotely (Garreau 2006). This military merger of human and machine is a fast-developing example of more broadly emerging **cyborgs**, the fusion of **cybernetics** and organisms, which also includes more mundane **prosthetic** extensions of human capacities such as pacemakers and contact lenses.

Anthropology is the study of humanity, but what humans are is in many ways being transformed in the present world. The close relationship many have with their cellphone is one everyday example of this: a survey found that most people would rather lose their wallet than their phone. Technology not only changes our interaction with the world and with other people, it appears that it is also changing our brains. We offload our memory to written documents, our phones, or digital assistants. Research has found that people who rely on GPS navigation have weaker abilities to form cognitive maps of their surroundings. Future technology, dramatized in movies like *Transcendence*, where human minds are uploaded to computer systems, could take this transformation of humanity much further.

For many, **posthumanism** is mostly about how new technologies are changing what it means to be human. The snippets above, and many more examples that follow, demonstrate the deep importance of these changes, and the complexity involved in making sense of their implications. Some people, often known as **transhumanists**, are enthusiastic about the possibilities for futures beyond the merely human. Some are seeking ways to merge themselves with machines in pursuit of enhanced capabilities. (Others pursue body modifications to achieve what they see as their trans-species identity.) At some point, those body–technology mergers could take them far beyond humanity as we now know it. These trends, both those currently being realized and those still only objects of imagination, have great implications for anthropology, with its focus on humanity. Anthropology offers powerful perspectives and tools with which to examine such trends, the focus of Chapter 4. Becoming transhuman through integration with machines, though, is only one part of the subject matter we consider in this book. Being human, we will argue, has always involved more-than-human elements, such as the microbiome, tools, and language. All of these more-than-human characteristics of humanity—past, present, and future—are concerns for interdisciplinary posthumanist research.

Our approach to posthumanism differs from those that concentrate only on the ways in which contemporary and future technologies make it possible to transcend conventional human nature. By contrast, in this book we adopt the somewhat paradoxical claim that we have always been posthuman. By this claim we mean that, much more so than any other animal, becoming human involved our intimate interaction with more-than-human elements. Biological anthropologist Richard Wrangham (2009) has argued that the control of fire and cooking made us human. Rob Dunn (2011) concludes that without certain bacteria, it would not have been possible for humans to survive on the agriculture-based diet that we adopted in the last 10,000 years. Without non-humans, there would have been no humans, because our nature is tied up with them for as far back as we can trace humans. Before technologies like cooking and language, we were not yet human in a very basic sense.

One way of differentiating transhumanism and posthumanism is that transhumanism is future oriented. It concerns things that are only just on the cusp of being possible, that excite the imagination or stir deep fears about the consequences of adopting new technologies for ourselves, cultures, and societies. Posthumanism, by contrast, is more inclusive and encompassing: it addresses all of the ways in which we have been more than human throughout the history and prehistory of the human species. One implication of posthumanist perspectives is that the earliest hominids who adopted fire and tools, or who used language to coordinate their collective subsistence activities, were engaged in what we might call a kind of ancient transhumanism. Adopting these new practices and tools, after all, did result in the transmutation of the characteristics of humans in ways even more revolutionary than the effects of contemporary communications technology. However, once these new practices are taken for granted or normalized, they become seen as simply part of the features of humanity. This is what we mean by the statement that we have always been posthuman: becoming human involved the adoption of new extrasomatic technologies (i.e., things that go beyond our bodies and their basic abilities) and fundamental changes in our microbial ecologies. Without these more-than-human extensions of our capabilities, we would not have become a species capable of dramatically changing the entire globe. Inhabiting the globe required collaboration with plants and animals; we have been tangled up with them since the beginning.

Scientists have proclaimed that we no longer live in the geologic era of the Holocene, but have entered the **Anthropocene**: the newly identified geological period in which human activity has become the dominant influence on climate and geological processes. One of the ironies addressed by this book is that just at the time when humans have developed the capabilities to become one of the dominant forces shaping the world itself, we need to become less **anthropocentric**. If we do not become more aware of the non-humans with which we share the world both inside and outside our skins, saving the world, ours and theirs, for the twenty-second century will be much more difficult.

The imperative to go beyond humans in our search for an understanding of humanity and its future can be seen as an extension of the anthropological mission to challenge **ethnocentrism**. Ethnocentrism is belief in the superiority of one's own culture. In research terms, though, the more damaging feature is the accompanying tendency to view other cultures from the perspective of one's own. This can often lead to misinterpretations, as we observe what they do and hear what they say through the lens of our cultural baggage. Distortions arise when we interpret the actions and statements of others in terms of what they would mean in our own culture. Cross-cultural conflict is often heightened by misunderstandings, which are inevitable if we don't become aware of our ethnocentrism. The difficulty of resolving the misunderstandings are obviously much greater when there is no shared language, as is the case in our relations with non-human animals.

Posthuman-ism or Post-humanism?

The other crucial difference between posthumanism as we are discussing it here and transhumanism is that posthumanism is just as much about rejecting and transcending the explanatory adequacy of *humanism* as it is about being posthuman. Our concern in this book is not primarily about *posthuman-ism*, the study of posthumans, those who have and will use technology to go beyond the "ordinary," or "species typical," human lifeways. It is much more generally and inclusively about *post-humanism*, about recognizing the biases and baggage of Western humanist worldviews. To be posthumanist, anthropology must reject anthropocentrism, the assumption that everything revolves around us humans. As geocentrism once assumed that the heavens revolved around the earth, anthropocentrism presumes that we are the main or exclusive source of momentum, dynamism, and value in the world, that we are the only species to intentionally set out to change the world and the one in relation to which those changes should be evaluated.

Modern **humanism**, as opposed to its early sprouts in the ancient world, is usually traced back to the fourteenth-century Italian Renaissance (Goody 2010), although Charles Nauert (1995) argues that leading humanists of the period, such as Petrarch, were still deeply religious. For our purposes in this volume, we find it most useful to view humanism as one of the fundamental elements of Enlightenment worldviews, which emphasize ideas of secularism, rationality, and the possibility of human progress and improvement. Rather than being at the mercy of supernatural beings, Enlightenment scholars argued, the fate of humanity was in its own hands. By rejecting medieval superstitions and constraints on free thought, humans could apply reason to understanding the universe. They could then transform it in ways intended to foster human welfare and other goals.

Posthumanist scholars have in recent decades criticized the humanities and social sciences for their excessive focus on the human, pointing out that it is a category that cannot be understood except by reference to the non-human (Wolfe 2010). The humanist attitude relies on the humanity–animality dichotomy, defining each in terms of the other. Becoming human is thought to be "achieved by escaping or repressing not just [our] animal origins in nature, the biological, and the evolutionary, but more generally by transcending the bonds of materiality and embodiment altogether" (xv). The usual way of thinking about the category of the human involves sharply separating us from animals. Yet in biological terms we are also animals, and we share with these others most of our physical nature. Humanists must also distinguish us from tools and technology, even though we are increasingly having tools implanted in us and are otherwise deeply dependent on them to do the things that make us human. Tools have made it possible to see ourselves as superior to, as well as fundamentally different from, non-human animals. The character of interactions between humans and other animals is the subject of a rapidly growing number of studies, in anthropology and other fields, the focus of Chapter 3. Often referred to as the study of human–animal relations or anthrozoology (Hurn 2012), these labels are seen as too anthropocentric by many, since they center humans in the naming of the field. A useful alternative is what has become widely referred to as **multispecies ethnography** (Kirksey 2014), which we adopt as the title for Chapter 3 (Smart 2014). Another non-anthropocentric alternative label is critical animal studies (DeMello 2012), although it excludes plants, unlike multispecies ethnography.

The ideas of progress developed in the Enlightenment ultimately asserted the perfectibility of both the individual and of society. It was necessary that irrational prejudices not get in the way of rational choices, which ideally would be based on scientific investigation of the problems to be solved. Transhumanism can be seen as humanist rather than posthumanist in this sense: technological enhancements and transformations of the "natural" human are consistent with emphasis on the improvement and perfectibility of the individual, although by means unimaginable by Enlightenment scholars. An example of this trend can be seen in the adoption of sensors and monitoring apps to collect information on daily activity and calorie input to achieve fitness goals, including to be "the best that you can be." In the first quarter of 2014, 2.7 million fitness trackers were bought, and by mid-2015, Fitbit alone had 20 million registered users.

The reader might ask, What is wrong with humanism? Doesn't it express the highest values and aspirations of civilization? Don't we want to be humane rather than inhumane? The question is more complicated than this. Historically, European humanism saw some people (women, people of other races, members of colonized societies, the working class, those with disabilities) as less human than others, or not human at all. Tony Davies (1997: 141) insists that "all Humanisms, until now, have been imperial. They speak of the human in the accents and the interests of a class, a sex, a race, a genome. Their embrace suffocates those whom it does not ignore. . . . It is almost impossible to think of a crime that has not been committed in the name of humanity." Feminists and postcolonial scholars have criticized humanism for its sexist and racist exclusions, but often do not reject it altogether, seeing the value of a less hypocritical humanism that includes all people. This position makes posthumanist perspectives ethically and intellectually troubling for them. Frantz Fanon "wanted to rescue Humanism from its European perpetuators arguing that we have betrayed and misused the humanist ideal" (quoted in Braidotti 2013: 24). Such critiques are useful for our purpose here because they reveal alternative ways to look at humanity more inclusively. Humanism does not accurately describe or explain the real human condition, our dependence on non-humans, past, present, or future. Even when stripped of its Eurocentric biases, as humanism must be to be useful for contemporary anthropology or social theory, it narrows our vision in ways that hinder our understanding of the contemporary world, or how we got here. We have to take one step further in challenging ethnocentrism, by considering how anthropocentric assumptions both mislead us and contribute to current world problems. The idea of the Anthropocene as the newest geological era should not confirm our anthropocentric bias, but rather highlight the many ways in which industrial and postindustrial humanity is intimately bound up with non-human co-travelers on our shared planet. Our survival may depend on understanding the company we keep, within our bodies, attached to our bodies, and around the world.

Anthropology and Holism without Boundaries

In this book, we introduce this broader concept of posthumanism, which includes but goes far beyond the science-fiction-becoming-reality of transhumanism. We believe, and hope to convince you, that the posthumanist approach helps us to better understand the human condition. A useful contemporary definition of anthropology is "the discipline that studies the whole range of diversity of ways of being human, by including in our collective research humans in all the times and places in which we have existed." Adopting a posthumanist approach can further this project and help us to transcend the flaws of the less inclusionary beginnings of our discipline.

Modern anthropology began in the context of nineteenth-century imperialism. Social and cultural anthropology adopted the task, in the intellectual division of labor, of studying the "primitive societies or cultures" being colonized by the countries where the earliest anthropologists were developing the discipline: Britain, France, Germany, the United States, and so on. Those Western societies would in turn be studied by the discipline of sociology. Since the beginning of decolonization after World War II, anthropologists have rejected the racism of this academic division of labor. We had to fundamentally reinvent our discipline, starting with the very definition of the field, which became simply "the study of humanity." Anthropology has expanded its disciplinary mission to explore the diversity of ways of being human across the entire time and space that we have been around. The expansion of our disciplinary scope continues. Many graduate students in anthropology study sites never considered in classic anthropology: cities, medical doctors, research laboratories, government agencies, and non-governmental organizations (NGOs).

To continue moving forward to achieve the ambitious mission of studying the whole range of ways of being human, we have to reinvent the classic anthropological principle of **holism** for a new age (Otto and Bubandt 2010). We need to pursue holism not only across borders, as in the anthropology of globality or transnationalism (Marcus 2010), but also across boundaries, importantly but not exclusively referring to species boundaries.

Holism is the principle held by most anthropologists, unlike practitioners of more specialized disciplines such as political science or economics, that we cannot usefully focus on only one aspect of a society or culture. This principle was said to apply because all of the parts of a society are connected and bound up with each other, which created a need to understand how the parts fit together in a particular context. From a holistic position, a "phenomenon has meaning, function, and relevance only within a larger context" (Otto and Bubandt 2010: 1). Research by Malinowski and others established that any social institution cannot be adequately understood in isolation from the other institutions with which it is entangled. This claim is particularly, but not exclusively, true for what we now call small-scale, stateless, or face-to-face societies—groups like the Trobrianders, Inuit, or Nuer—studied in classic ethnographies. It would be artificial to separate out politics from kinship or economics among the Trobrianders, for example. Power was achieved through birth into hereditary positions. The wealth needed for the chief to put on feasts, a key source of support for his prestige, was accumulated through marrying multiple wives, whose brothers gave most of their harvest to the chief.

The complex global entanglements of the contemporary world mean that to understand local societies, we not only have to consider interactions with people outside that society, but also have to address our non-human co-travelers on this planetary journey: microbes, parasites, domesticated species, and technologies. In this task, we can usefully return to the holism of classic ethnographies in which cattle, pigs, yams, sorcerers, and ghosts were central to the lives and livelihoods studied by anthropologists. To be holistic in the contemporary world complicates the task of classic holism, and not only because dislocating it from a presumably bounded local society means that a fully holistic study is impractical, if not impossible. More fundamentally, transcending the research focus on a locally delimited group makes the distinction between parts and whole unworkable: What is the whole within which a part must be understood, if a society is bound up for important purposes with the entire world? The context that is important will vary by issue and by time, so that a holism without boundaries can never be more than an incomplete project to study the context that emerges through research that is important for the issue of interest. We suggest that a practical partial solution to this problem is to follow up the connections that emerge as needing to be considered. We can best do this through an open-ended ethnographic exploration. We believe that one of the crucial messages from posthumanism is that our ethnographic efforts must pay attention to entanglements between non-humans and humans, as well as between people in different locations. The relationship between any organism and its environment can be better understood as entangled rather than bounded. The "environment" is not something outside a bounded entity, but a "zone of entanglement" within which beings grow "along the lines of their relationships" (Ingold 2008: 1807). In other words, our ties to other people, things, and places influence the paths that we choose to follow. Following the paths that emerge from more-than-human relationships can lead us, and the discipline of anthropology, to strange and exciting vistas of futures that may be in the process of coming into being.

An example illustrates the complexity and unpredictability of entanglement. Hong Kong helped China, after economic reforms began in 1979, to become increasingly prosperous (Smart and Smart 2016). In Guangdong province, this encouraged the expansion of the expensive practice of eating wild meat. These culinary choices provided the channel for the jump of the new infectious disease severe acute respiratory syndrome (SARS) from civet cats to humans. The intense economic and social integration between Hong Kong and the rest of China meant that SARS quickly moved to Hong Kong in 2003. It struck Hong Kong's economy a deep blow. China tried to help Hong Kong, and one measure was to allow more mainland tourists to visit Hong Kong. This Individual Visit Scheme meant shopping expenditure by mainland Chinese soared from HK$19.9 billion in 1999 to HK$158 billion in 2012. The share of mainlanders in the total tourist numbers for Hong Kong grew from 33 to 58 per cent. This situation, in turn, contributed to the rise of an anti-mainlander movement in Hong Kong. Such situations reveal a tangled web of connections, the effective management of which is extremely difficult. New zoonotic diseases result from human expansion into and disruption of ecologies. Human expansion is creating not only a wave of extinctions but also a need for the parasites of those endangered species to find new hosts and ecological niches. Their adaptation to our invasion can spread catastrophes across many populations, not only human ones.

To equip anthropologists to incorporate microbiomes, the Anthropocene, and implanted **nanotechnology** into our toolkit, we need to recover and rethink classic holism. In doing so, we need to extend it into social and cultural worlds that are neither locally bounded nor exclusive of the non-humans that are bound up with human ways of life. At present, but also in the past, these newly emerging connections trace spectacular and sometimes terrifying trajectories into futures, rushing forward at an intensifying pace. For example, cheap 3D printers are allowing insurgents to produce more deadly improvised explosive devices (IEDs) by downloading instructions. They also allow people to print their own dental braces and prosthetic limbs. For clues to what might happen in the near future, stay tuned to not only the scientific literature but also science fiction novels and films, which often add to wish lists for technologists who try to bring visions of **cyberspace** or tricorders into workable reality.

This short volume will lay out some of the interesting insights that we can obtain by following the methodological principle of *holism without boundaries*. Our firm belief is that by continuing in this direction, anthropology will add ever more to our collective knowledge of more-than-human nature in past, present, and future, and the consequences of becoming the kind of people that will emerge in the coming decades. We must make clear, though, that we find any single discipline, including anthropology, inadequate to the demands of actually conducting holism without boundaries. As humans collaborate with microbial fellow travelers, tools, and companion animals, we must also collaborate across the narrow, parochial divisions within academia. To understand what posthumanism can offer to our understanding of past, present, and future worlds, we cannot restrict our vision within narrow disciplinary definitions. Instead, we need to cast our nets widely, and learn from a broad and cosmopolitan community of scholars who are working to better understand our more-than-human entanglements. Yet, it is also important to recognize where we come from, and in reconsidering the scholarly paths we have followed, to display the distinctive contributions that those influenced by an anthropological tradition of holism might be able to make. For this, we must occasionally turn to our discipline's history.

In anthropology's traditional domain of the study of small-scale, non-Western societies, we did not have to carve out a specialization, a focus on a particular part of the social and cultural world, within a division of labor with other researchers working on the same societies. The earliest anthropologists who formed the field were invariably the first to study their societies in any depth. The principle of holism encouraged us to consider the associations between humans and non-humans (pigs, ancestors, ghosts, medicine, poison, totems, reindeer, forests, and so on) in ways that most sociologists were not inclined to do. Most sociologists sought to demonstrate the importance of "the social" in relation to distinct domains of nature and the economy. The local holism of anthropology, however, was made possible only by the artificial bounding of societies or cultures (Leach 1976). To build on its advantages, anthropology needs to follow associations beyond our comfort zones and to accept the challenge of learning from our entanglements across all kinds of boundaries. Anna Tsing (2010) has argued for such an opened-up version of holism, and describes her book on matsutake mushrooms as sketches of "open-ended assemblages of entangled ways of life" (viii).

Posthumanism: Fad or Productive Approach?

Some readers may by this point be feeling dubious about the idea of posthumanism. They might ask, for example, if it isn't just another trendy member of the "posts" (postindustrialism, postmodernism, post-structuralism, postcolonialism). If so, they are right to be skeptical, since that is always a good starting position for any intellectual inquiry, including reading a textbook like this one. We had such concerns when we first started reading seriously in this area. The ideas seemed fascinating and relevant to our research at the time, but we worried that it was mostly a matter of playing with words to appear to have something new to say, and thereby hoping for the attention of other readers and those who hire and fund researchers. It was only when we wrote a book chapter on the movement of non-human life across borders that we became convinced of posthumanism's importance (Smart and Smart 2012a). We discovered that almost all of the research on borders concentrated on humans. Once we read extensively beyond the field of "border studies," however, we discovered clear evidence that for centuries non-human life has had a huge impact on the management of borders. Its impact can be traced to life's ability to replicate itself. Self-replication is particularly problematic in the form of epidemics like the Black Death. Life makes the management of borders difficult, because many forms of life can move by themselves or through their seeds. Mobility can produce problems of invasive species, the control of which costs billions every year. Alberta devotes resources to keep that province rat-free. The problems caused by rabbits in Australia are notorious. Keep your eyes out for road signs that encourage careful scrutiny against invasive species, such as those in Canada that warn boaters to clean their boats, so as not to bring in Eurasian watermilfoil tangled around their propellers. We return to invasive species in Chapter 3.

When we considered what these omissions and gaps in our knowledge about borders might tell us, we realized that anthropocentrism had operated as a set of blinders that prevented balanced attention to the major influence of non-human life on national borders throughout history. Compared to non-living materials, the challenge posed by living organisms lies particularly in their mobility and ability to reproduce, often at spectacular rates when conditions allow, like the rabbits introduced to Australia. Artificial intelligence and robots, however, challenge this kind of distinction, as seen in the worries that self-replicating nanobots gone wild might turn the earth into nothing but gray goo. Movies like *The Matrix* and the *Terminator* series portray the nightmares of futures where computers take over. We return to such issues in Chapter 4.

Borders were important to us because we were once engaged in research on the impact of the 2003 outbreak of "mad cow disease" or BSE in Canada. It caused an economic catastrophe for Alberta farmers, but resulted from microscopic agents: infectious proteins from cattle that could cause variant Creutzfeld-Jakob disease, an incurable and always lethal disease of the brain in humans. These proteins, prions as they became known, changed the world in many ways. Global trade regulations, food safety rules, the entire meatpacking industry, and many other institutions were modified in profound and consequential ways. How can it be that something that could be seen only under a powerful microscope, with fewer than 30 cases in Canada, had such far-reaching impact? From our own experience, doing research on BSE forced us to move far from the areas in which we had developed expertise, particularly with food producers and consumers, and learn about prion science and global policy making. Holism without boundaries was not something we chose to pursue. Rather, it was something that emerged from our efforts to understand what we were seeing in the field, literally in the fields of Albertan cattle producers, but also elsewhere: in government offices, in the meetings and discussions of NGOs like the Alberta Beef Producers, and in urban disputes over new slaughterhouses intended to help beef producers suffering from the consequences of BSE.

Organization of the Book

In the chapters that follow, we will concentrate on three dimensions of posthumanism that we find particularly fascinating and relevant for anthropology. The next chapter looks within, to the microbiome, the diverse universe of microscopic creatures that live within and on the outside of our bodies. It also looks outward, to the global impact of zoonotic diseases, those that can be transmitted between animals and humans, such as bubonic plague, BSE, avian flu, swine flu, and anthrax. Epidemics have had a massive impact on human history, and research now suggests that most human diseases had their start as zoonotic diseases. In addition, the microbiome is one of the hottest areas of research into ways to combat disease and improve human well-being, and as a result is also one of the newest investment boom fields. Our examination will thus move through the whole range of scales of analysis, from microscopic to global. Our movement outward is similar to the approach in Hinchliffe and Woodward's 2004 book *The Natural and the Social*, which they say is organized to move from inner nature, "bodies, personalities, and what makes us human," to outer nature, "the environments and the non-human world within which we live and coexist with other people, plants and animals," to break down the "assumed boundaries between nature and society" (Hinchliffe and Woodward 2004: 1). These separations between inner and outer are themselves problematic, however, in the holistic perspective of a world where everything is tangled up with everything else in one way or another, even if only through our mutual involvement with climate change. The world does not end at the surface of our skin, which connects us to the outside as much as it separates us from it. Nor is everything that makes us human inside us. The chapter will also introduce some key concepts from ethnographic studies of science that will be used throughout the remainder of the book.

Chapter 3 turns to the study of humans and their interactions with (multicellular) animals and plants. Ideas of humanity are commonly tied up with distinctions from animality, but the interactions between humans and other animals, and the understanding of such interactions, vary substantially from one culture to another. Anthropological study of our relations with non-human animals commonly attempts to characterize the different ways in which such relationships are culturally organized, and to discover how much variation exists among the world's many societies. Another goal is to consider the effects of those diverse cultural systems on our interactions and their consequences, particularly but not exclusively, for our economies and ecologies. But these kinds of approach, valuable as they are, still begin from human understandings of non-humans, and create problems for our goals here. Going beyond Western ideas of the human–animal difference is an indispensable task if we want to avoid ethnocentrism, but by concentrating only on how different groups of people think about animals, we remain within an anthropocentric framework. How we go beyond anthropocentrism is a central issue of this book.

Many posthumanists reject most or all divisions between humans and animals. This position can be understood as a political one, but a direct political expression of it such as in animal rights activism is only one expression of such ideas, although a very important one. For many other researchers, breaking apart systems of thought that are based on unsupportable dualistic ideas is also crucial work, which can help to provide cultural foundations for an attack on **speciesism** (an idea we consider in Chapter 3). Several anthropologists have been arguing that ideas of **multiculturalism** are too weak to express differences in this realm, and argue that cross-culturally there are diverse forms of **multinaturalism**. That is, they argue that humans live in radically different natures rather than merely having different worldviews and understandings of nature in the Amazon or the wheat fields in Saskatchewan or the environment of Toronto. We suggest that classic ethnographies have something to offer to a post-anthropocentric anthropology in the inclusive holism across boundaries that included important animals, plants, substances, and places.

Chapter 4 considers transhumanism as the enhancement of human capacities, but by looking backward as well as forward. We first consider the earliest prehistory and history of prosthetic extensions of human abilities, which have been argued to have been key parts of the process in which we became human. We then examine the implications of technologies that are already widely used. Finally we turn to more speculative discussions of the potential for radical transformation of the nature of humanity in the near and longer-term future. A discussion of science fiction novels and films is useful because of the familiarity of some of their sources, and the ways in which insightful authors such as William Gibson have helped to create the future through the naming of aspirational technologies such as cyberspace. Religion must also be included in a non-anthropocentric perspective, since around the world it serves as an important conceptual lens for people trying to understand new worlds and the spiritual, moral, and practical tangles that they often generate. We consider at some length discussions by theologians about transhumanism and its meaning for spiritual belief. After all, ghosts, ancestors, and witches are as much a part of traditional ethnographies as yams and cattle. Our considerations draw on a rich stew of anthropological research by scholars in a range of disciplines, from archaeology to social anthropology, and many others who engage in either ethnography or science studies. Several anthropological studies of in vitro fertilization (IVF) are used to draw out some of the implications for classic anthropological topics such as kinship, exchange, identity, and nature, as for the first time in history there are people alive who would not exist if their births had not been mediated by sophisticated technologies that allow reproduction to begin outside the biological womb.

Posthumanism is not limited to the three important subtopics addressed in these chapters. There are many, too many, issues that arise: the theoretical sources and debates within this field; the relationship between anthropological and other approaches to posthumanism; the formation of posthuman subjectivities; the incorporation of non-humans into political decision making; and the implications of posthumanism for policy responses to contemporary crises, to cite only a few examples.

Our main hope is that the next three chapters, despite their brevity and lack of comprehensive coverage, will introduce the reader to the approach and to a wealth of insights from writers who open our eyes to fascinating new ways of seeing the worlds in which we live. In doing so, they tell us riveting stories and lay out possibilities for what humans might become. We will also suggest some of the possibilities and advantages that open up when we pursue anthropology into new arenas by practicing holism without boundaries. When we trace the interactions beyond humanity to the non-human agents that have been involved in the formation of our past and present, we can see ourselves in new ways, not separate but entangled.

Readers will be able to observe some of the tremendous transmutations that result during their own lifetimes. Our hope as teachers is that readers will think about their own practices and situations using some of the ideas we explore in this short book.

Discussion and Activities

Discuss the idea of progress with your peers and older relatives or friends. What do you and they think of the idea? Have science and reason contributed to the improvement of our lives and our world? Do your attitudes toward the future differ from those of your older relatives or friends?

Identify an issue in a current news article that raises an issue relevant to posthumanism. How might ideas introduced in this chapter raise questions about that issue? Is the treatment of the issue in the article anthropocentric? If so, how does that influence the emphasis? What questions are obscured by the anthropocentric emphasis?

Compare the posthumanist approach introduced in this chapter with the other texts used in the course you are studying (if applicable).

Think of a situation in which you are entangled with non-humans. How far can you trace the connections? What kind of entities are bound up in these interactions, and how do they influence the interactions?

Watch a science fiction film, such as the ones listed below or in subsequent chapters, and discuss it, using concepts introduced in this chapter.

Additional Readings and Films

Note: These lists of readings at the end of each chapter are intended to be supplemental to the references cited, which in some cases may be more technical than desirable for undergraduate reading. In addition, they includes several key sources on issues that we have not engaged with at any length here, such as the ethics of animal rights, which may be of particular interest to students grappling with the meaning of the debates we have addressed for their own ethical positions and commitments.

Readings

Braidotti, Rosi. 2013. *The Posthuman.* Cambridge: Polity Press.

Haraway, Donna. 1991. *Simians, Cyborgs and Women: The Reinvention of Nature.* New York: Routledge.

Hayles, N. Katherine. 1999. *How We Became Posthuman: Virtual Bodies in Cybernetics, Literature, and Informatics.* Chicago: University of Chicago Press.

Hodder, Ian. 2014. "The Entanglements of Humans and Things: A Long-Term View." *New Literary History* 45 (1): 19–36.

Ingold, Tim. 2000. *The Perception of the Environment: Essays on Livelihood, Dwelling and Skill.* London: Routledge.

Knauft, Bruce M. 1996. *Genealogies for the Present in Cultural Anthropology.* New York: Routledge.

Lorimer, Jamie. 2015. *Wildlife in the Anthropocene.* Minneapolis: University of Minnesota Press.

Nading, Alex M. 2014. *Mosquito Trails: Ecology, Health, and the Politics of Entanglement.* Berkeley: University of California Press.

Wolfe, Cary. 2010. *What Is Posthumanism?* Minneapolis: University of Minnesota Press.

Films & TV Series

Battlestar Galactica. 2004–2009. TV series directed by Ronald D. Moore. Los Angeles: David Eick Productions.

Rise of the Planet of the Apes. 2011. Film directed by Rupert Wyatt. Los Angeles: Chernin Entertainment.

Transcendence. 2014. Film directed by Wally Pfister. Los Angeles: Alcon Entertainment.

Twelve Monkeys. 1995. Film directed by Terry Gilliam. Los Angeles: Atlas Entertainment.

CHAPTER 2

ZOONOTIC DISEASES AND THE MICROBIOME

When we cross a border, we present a passport, but we don't usually declare the horde of microscopic "companion species" that accompany us inside and on the surfaces of our body (Haraway 2008). On occasion, we are asked to declare if we have been exposed to places or people with diseases of current concern, such as SARS, avian flu, Ebola, Middle East Respiratory Syndrome, or Zika virus. On many customs declaration forms, there is a box to fill out to indicate if you have been on a farm while outside the country and will be going to one in your home country. These acknowledge the existence of what has come to be known as the microbiome, one subset of which includes **zoonotic diseases** (those which can be transferred between animals and humans). Such zoonoses are now seen as a major source of security risks. Managing such risks has come to be called **biosecurity**. The **microbiome** has also become a rich arena for efforts to improve the health and well-being of humans by influencing the composition of that internal ecology rather than simply eradicating "harmful germs." Manipulating the microbiomes of other organisms is a rapidly growing field of research. The coffee berry borer is one of the worst threats to coffee beans ("Beetles and Bugs: Protecting Coffee Crops" 2015: 68). Unlike most other insects, for which the caffeine is toxic, this borer has gut bacteria that eat the caffeine before ill effects occur. Researchers are exploring ways to attack the borer's bacteria rather than the insect, with hopes that this would reduce negative environmental side effects. For both our own health and that of our crops, intervening indirectly through the microbiome rather than directly on the targeted species might offer advantages. In this chapter, we consider first the human microbiome. Then we use some posthumanist ideas to discuss our complex entanglement with microbes. Finally we consider zoonotic diseases and the global impact of microscopic entities.

The Microbiome

Some of our "tiny companions" (Haraway 2008) are parasitic, even potentially deadly, while others contribute to our bodily functions, such as efficient digestion and absorption of essential minerals, as well as our immune system. The microbiota in our digestive system provide "an indispensable internal ecosystem for numerous host physiological processes and can be considered to have coevolved with the host to form a superorganism" (Honda and Littman 2012: 761). Hosting a constellation of invisible life forms, as well as being dependent on them in a variety of ways, seems far from the autonomous, world-improving individual of humanism. Those thinking that humans are carefully, and best, divided from all other creatures, of which they are sovereigns or stewards, may find their skin literally crawling as we consider the way in which human life is tied up with things in our body that we cannot even detect without technologies that magnify our observational capacities.

Perhaps the solution is to thoroughly cleanse our body of such invaders? This was certainly a powerful theme at the peak of progress thinking. The advertising and pharmaceutical industries sold us, and continue to sell us, a corporate fantasy of eradicating all microbes through household cleansers, mouthwash, hand sanitizers, antibiotics, ear cleaners, and other products. Yet this sanitized utopia is unlikely to be achievable, even if it were desirable. Our temporarily successful interventions against bacteria usually have the result of breeding more dangerous strains, and the more antibacterial products we use, the greater the risk. Even if an attempt to get rid of all microbes could succeed, it would probably be dangerous, due to the various dependencies humans have within our microbiome. The experiments of James Reyniers, which helped to nurture the microbe-free dream, are instructive.

Reyniers, born in 1908, was fascinated by Louis Pasteur's work and wanted to know if it would be possible to remove all the bacteria from an animal or human. A machinist and 19-year-old undergraduate at Notre Dame University in Indiana, rather than a biologist, he took the approach of adapting the recently invented iron lung to construct a microbe-free chamber and have mothers give birth in that space. Surprisingly, a dean gave him space, metal, and tools to make the attempt. In 1935 he finally produced the first generation of apparently germ-free guinea pigs. Rather than dying without the microbes, as Louis Pasteur had speculated, they survived and had better appetites than other guinea pigs. His results were widely taken up by the popular media and used to support the idea that a germ-free world would be one in which we were healthier. The problem was, however, that the guinea pigs were not bacteria-free, since some necessary species were being passed from mother to offspring directly (Dunn 2011). In addition, "germ-free" guinea pigs needed to be fed extra food that was richer in nutrients to gain the same amount of weight, which would hardly be an advantage for animals in the real world (although in our contemporary world, the weight loss industry and its customers would probably find this feature attractive).

Most organisms are occupied by microbes with benefits. Some microbes provide enzymes that their hosts lack, and that allow their hosts to use a larger proportion of the nutrients in their food, particularly from plant material. *Bacteroides thetaiotaomicron*, for example, common in human guts, produces over 400 enzymes that we otherwise lack for breaking up plant material. When food is scarce, our "microbes make it less so. The microbes in … our guts, produce up to 30 per cent more calories from food than the hosts can produce on their own" (Dunn 2011: 79). Truly germ-free mammals cannot produce or acquire enough vitamin K from breast milk. Babies are at risk of death from hemorrhagic disease of the newborn without vitamin K, so they are routinely given a shot of it at birth. In countries where this is not practiced, hemorrhagic disease is more common in babies delivered by caesarean section. Rob Dunn (2011) argues that bacteria

> provided vitamin K where it was once scarce, but just as importantly, they allowed us to extract extra calories from our food, up to 30 per cent extra. More of those calories would, in turn, have been converted to fat on our bodies, which, for most of our history, was a good thing. In other words, they were our mutualist partners. Most years, but particularly the lean years, their offerings would be the difference between life and death. Most years in our history, we would have survived by dint of our microbes. If one had to spend ten hours a day gathering food without microbes, the gathering day was shortened to seven or even just six hours with microbes. (81)

What was evolutionarily advantageous, however, may now contribute to the "epidemic of obesity." Manipulation of our microbiomes is being researched for possible responses to health problems related to obesity, where chronic "lifestyle diseases" (cardiovascular disease, diabetes, etc.) are replacing infections as the major health problems, in the emerging economies as well as the rich world. Some research suggests a positive feedback loop where a diet inducing obesity changes the gut's microbiota, leading to more efficient extraction of energy from the diet, helping in turn to perpetuate obesity (Honda and Littman 2012). As in the case of all ecologies, the relationships are complicated. Recent research suggests that artificial sweeteners actually increase the risk of obesity due to negative impacts on gut bacteria that result in glucose intolerance, a major risk factor for diabetes as well ("Saccharin Solution?" 2014).

In 2007, the US National Institutes of Health established the Human Microbiome Project (HMP) with a five-year budget of US$115 million. This "imposing yet logical conceptual and experimental extension of the human genome project ... promises to break down the artificial barriers between medical and environmental microbiology." It is intended to "define the parameters needed to design, implement and monitor strategies for intentionally manipulating the human microbiota, to optimize its performance in the context of an individual's physiology" (Turnbaugh et al. 2007: 804).

Just as the human genome project helped to launch the biotechnology industry, the microbiome has become one of the hot medical and investment stories of the last few years. Seventure Partners has one of many investment funds targeting microbiome companies, with €100 million under management. Seventure's CEO is particularly interested in the potential to apply gene-editing techniques to the microbiome. She suggests that this approach could help address antimicrobial resistance. In 2015 Harris and Harris financial advisors identified the microbiome as one of their technologies that will drive innovation and investing. They write about pills "being developed to get good bacteria into our system to treat gastrointestinal issues. Our own portfolio company ... is actively involved in commercializing discoveries from the interaction of the microbiome with human health ... far more diseases than previously known may result from human interaction with the microbiome" (Jamison 2014).

Incorporation of non-human species within the enveloping skin of the human body[1] goes back further than our hominid ancestors and our **coevolving** bacteria. All living things, including humans, have "mosaics of genomes, with hundreds or thousands of genes imported by viruses." The history of genes is "like a bustling trade network, its webs reaching back billions of years" (Zimmer 2012: 45). The mitochrondria, essential organelles that release energy in the cells of all multicellular creatures, were originally bacteria that became incorporated into cells. They are inherited exclusively from the maternal line, which has helped us trace human evolutionary genetics. Donna Haraway (2008: 31) describes the basic story of evolution as one in which more complex life forms are the "result of ever more intricate and multidirectional acts of association of and with other life forms."

---

1 Tim Ingold (2011: 87) says convincingly that our skin connects us to the world more than it separates us from it.

Beyond the physical body alone, evidence is emerging that some microorganisms and parasites can affect our emotions and actions, not just our physical health. Research on germ-free mice found that a complete absence of microbiota led to decreased anxiety-like behavior and caused changes in neurochemistry (Cryan and O'Mahony 2011). One example may help to outline the implications.

Toxoplasma gondii is a protozoan, acquired by exposure to cats or certain foods. Latent infections, without apparent symptoms, affect 30 to 50 per cent of the world's population. Active cases of toxoplasmosis can result in brain **parasitism**, and can influence personality and behavior in humans as well as in other animals (Flegr 2007). Chronic infections are associated with behavioral changes that are known as "host behavior manipulation." Infected rats and mice prefer exposed areas in unfamiliar environments, stand up more often, and generally express fewer signs of fear. They are more likely to be victims of cats, which would help transmit the parasites (da Silva and Langoni 2009). Chronic human cases have been reported to show unusual gender differences, with one study finding that men display suspiciousness, shorter tempers, and disregard for social rules, while women seemed more "warm hearted," conscientious, and moralistic (Flegr 2007). Researchers have suggested that **schizophrenia** is related to toxoplasmosis: schizophrenics are more likely to show chronic infection (da Silva and Langoni 2009). There are other examples of parasites changing behavior, perhaps the most spectacular being a fungus that produces "zombie ants" which take a death grip on a branch tip until they die, at the right time and place to assist parasite reproduction (Hughes et al. 2011).

One of the most important contemporary trends with regard to the microbiome is the way in which it is increasingly being viewed as a set of tools that we can use to solve bodily problems and improve human lives. This is a marked shift from the twentieth century, when "germs" were seen as scourges to be eradicated as much as possible, based on the belief that an abiotic environment would be a safe environment. We can see the beginning of a shift toward public acceptance, even enthusiasm, for ingesting microorganisms to improve digestion and other issues, in the form of **probiotics** in yogurt and other products. The extent and speed of this shift can be seen in a simple Google search, which finds 13,200,000 hits for probiotics. A Google Scholar search finds 169,000 hits, 15,300 for 2014 alone, compared to 4,470 in 2004 and only 425 in 1994. Rapid growth of medical knowledge may soon allow probiotics to be tailored to individual needs based on an assessment of absences or other issues in one's personal microbiome.

Seeing "alien invaders" in our body as potentially beneficial rather than simply dangerous goes further. According to the Fecal Transplant Foundation, although fecal transplants were used in fourth-century China, and sporadically in the US since the 1950s, they have "gained popularity in the U.S. in the past few years."[2] In 2013, the Federal Drug Administration (FDA) classified fecal matter as both an Investigational New Drug (IND) and a Biologic, and ruled that "only physicians currently in possession of an approved IND application would be allowed to continue performing fecal transplant." Fewer than 20 American physicians were allowed to do it, either with fecal material inserted by colonoscopy or by the ingestion of pills, leading to "a groundswell of opposition from physicians and patients." The FDA reversed its position, and announced that qualified physicians could continue to perform fecal transfer for recurrent *Clostridium difficile* (a bacterial infection that in 2013 affected 347,000 Americans, of whom 14,000 died), only with signed consent from the patient.

Relatedly, there is a growing belief, still unproven and contested, that the home environment in the rich countries has become "too clean," resulting in a growth of allergies and autoimmune diseases. Perhaps the most extreme response to this has been the growth of a black market for helping people deliberately infect themselves with parasitic worms. The idea is that in the absence of predators to fight against, the immune system turns against its own body. By "rewilding" the gut, an ecological balance is reset within the microbiome. So far, these ideas are supported mostly by anecdotal evidence, but there is a growing amount of research being undertaken on the approach. Obviously there are good reasons not to have many kinds of parasites, so Rob Dunn (2011: 57) suggests that we may need "to domesticate our worms, to make their effects more predictable and their consequences more controlled." Parasites, bacteria, and other microbes, then, are increasingly seen as tools that can be used within our bodies, to trigger biological processes and mechanisms to achieve goals that in the past were left to pharmaceuticals and surgery.

Microscopic organisms may seem a long way from the transhumanist imaginings of minds uploaded to computers and technological augmentation of our bodies, or even from the conventional concerns of social and cultural anthropology. To show how they are related to the broader conception of posthumanism, we consider the ideas of several scholars here. These concepts will be relevant through the rest of this book.

Actor Network Theory: Things with Agency

Bruno Latour, who received degrees in both anthropology and sociology, is the best known proponent of **Actor–Network Theory (ANT)**, which originated in science and technology studies (STS) but has been widely

---

[2] "What Is FMT?" Fecal Transplant Foundation (2016), http://thefecaltransplantfoundation.org/what-is-fecal-transplant/.

adopted in other fields of research. Latour (2005) proposes rejecting sociology in its current form as the study of "society," to be replaced by the study of associations: all of those linkages that influence life, not just those linking us to other humans. This approach is posthumanist in the sense that it dissolves the dualism between humans and non-humans, concentrating instead on the web of relationships and interactions that make things and actions possible. Any action requires linkages to other networks, which we might also think of as chains of other actions. For example, "taking a course" at a university requires a wide variety of linkages between humans organized into institutions (the university as a whole; the registrar's office; student finance agencies; utilities providing electricity and water; forestry companies that turn trees into paper for textbooks; other companies that produce cement, steel, and buildings, etc.). In ANT, **actants** are entities, human or non-human, living or non-living, that make a difference in a controversy, action, or chain of events. In humanist perspectives, only humans have **agency**, the ability to intentionally make, or attempt to make, things change in particular ways. ANT avoids talking about agents, and uses the term *actant* to sidestep the assumptions in the more usual term *agent*. This distinction, though, has not kept ANT from being widely criticized for its lumping together of humans and non-humans.

The ways in which ANT scholars do their work has significant echoes with anthropological practices. As pointed out in the introduction, in early anthropology's traditional domain of the study of small-scale, non-Western societies, we did not have to divide up subject matter with other researchers. As a result of this situation, and also because so little was known about these small-scale societies at the beginning of the twentieth century, the ethnographers of the day usually tried to give a full account of the lives of the people they studied, in our terms, following up many of the entanglements with non-humans that affected their lives. Doing so led to much greater attention to animals, plants, and non-human elements of the landscape than in sociology, since these non-humans were so much a part of their everyday life and livelihoods (Latour 2005). Recall, though, that the holistic practices of these ethnographers was restricted to, and made possible by, what they conceived of as localized societies or cultures, deliberately neglecting extralocal connections such as colonialism and trade networks.

Latour (1993) has argued that modernity (by which he means more or less the same historical developments and characteristics that we describe as humanism in the introduction) was characterized by an unprecedented conceptual separation between humans and non-humans, between culture and nature. This dualistic separation was crucial to the emergence of modern science and capitalist economies. Scientific development and capitalist economic expansion, however, generated a massive expansion of technological capacities, thereby constantly producing human–non-human hybrids, such as manufacturing assembly lines, which integrate human labor with the pace set by the machines. For ANT, a hybrid is any entity that combines human arts and sciences with natural entities, thereby crossing the distinction that modernity put in place between human and non-human or culture and nature. Inoculations against diseases, through injection of killed microbes to promote antibodies, are an example of such a hybrid that creates new entities crossing this conceptual divide that modernity put in place. Another example is the climate change crisis wherein our fossil-fuel-driven machines produce carbon emissions that are changing the world's climate and which is now demanding changes in our economies and technologies to keep temperatures rising by less than 2°c, the target adopted in the December 2015 Paris Agreement by 193 nations.

Critics of ANT object to the equation of humans and other things as actants, entities possessing agency through their ability to make a difference within networks of activity. Such critics argue that "true" agency, the ability to make things happen, requires **intentionality**. Intentionality, in turn, requires consciousness and the capacity to plan and develop new approaches rather than simply relying on instinctual behaviors or material properties. The idea that non-living things might be equated with human agents is particularly objectionable for such critics.

If we accept the redefinition of what an actant is and does, these objections seem less convincing. For example, whether or not oil exists in sands in northern Alberta, the effect of its extraction on the environment depends on humans interacting with it: the motive force in the linkage between the humans and non-humans in this network comes from the humans. The distinction between actant, focused on by ANT, and agent, the emphasis of humanists, partially answers this criticism. ANT can concede that humans are different in significant ways from other actants, but see this distinction as unimportant for their intellectual project, since all actants differ from other kinds of entities in various ways. It is the ability to make a difference that is the main concern, not whether this is done by intent or through other mechanisms. In any case, the intentions of human actors always rely on recruiting and using non-humans in the networks of actions that may bring humans' intentions to fruition. Instead of prioritizing the allegedly distinctive intentional agency of humans, ANT takes an open-ended approach to explore what determines the development and transformation of particular entities and networks. What identifies an actant as important, what allows it to have consequences, is that its ability to "make a difference" in a particular chain of events is based on the specific features of that entity. If it had different properties, the network of action would either not come into being or would develop in another direction. For example, if the chemical bonding of oil to sand were either much weaker or much stronger, either a very different extraction industry would have developed

POSTHUMANISM

in northern Alberta, or perhaps it would not have developed at all. In a classic study, Michel Callon (1986) demonstrated how the specific features of scallop reproduction played a role in effecting the kinds of changes to a fishery that could make it sustainable, and that misunderstandings of scallop biology had previously resulted in the fishery's decline.

Eduardo Kohn (2013) sympathizes with the extension of agency to non-humans, but argues that a crucial distinction is being obscured with the concept of actant; that is, it rejects the fundamental difference between living beings and non-living objects. He argues that ANT's "approach to nonhuman agency overlooks the fact that some nonhumans, namely, those that are alive, are selves" (92). He is developing an "anthropology beyond the human," which is concerned with "exploring interactions, not with nonhumans generically—that is, treating objects, artifacts, and lives as equivalent entities—but with nonhuman living beings in terms of those distinctive characteristics that make them selves. Selves, not things, qualify as agents. Resistance is not the same as agency" (92). For Kohn, living non-humans have agency similar to that of humans, while non-living objects are, at best, actants due to the resistance created by their specific properties when other selves attempt to include them in their actions and networks. Kohn's objections are reasonable at one level, in that there are important differences between the drive to maintain the integrity, survival, and reproduction of living forms, whether unicellular, multicellular, or human, and the consequences in networks of action created by non-living materials.

From another perspective, however, there are a large number of other distinctions that might be made between different kinds of actants. ANT's reason for focusing on the similarities of actants by lumping them together is not to deny differences between them, but to make possible a research agenda that follows up connections and to understand the particular kinds of consequences that actants may have. ANT sees this as producing a symmetry of analysis: both humans and non-humans are analyzed using the same terms and methods rather than reproducing humanist dualism. The kinds of things that ANT and Kohn (2013) are trying to explain are simply different, and do not seem to us to be incompatible from a broader perspective. We can appreciate the advantages of both approaches by accepting the anti-dualist symmetry of focusing on actants, while recognizing that actants are extremely diverse, with some having capacities that others don't. In this line of thought, the ability of water to shift between solid, liquid, and gaseous forms without "dying" is a distinctive feature of a particular kind of actant equivalent in importance to the capacity of humans to plan out strategies to achieve goals in the future. Making distinctions within the universe of diverse kinds of actants is a different kind of intellectual endeavor, and the importance of Kohn's project doesn't mean that we need to reject the ANT approach.

Andrew Pickering (1995), a key figure in science studies, goes beyond ANT by acknowledging that intentionality makes an important difference between human agency and material agency. He says that it is necessary "to recognize that scientists usually work with some future destination in view, whereas it does not help at all to think about machines in the same way. Human intentionality, then, appears to have no counterpart in the material realm" (17). His position might be thought to be simply a return to humanist dualism, but it is more subtle than that. He argues that intentionality is only a small part of human agency. Because science, and any other complex human activity, relies on material actants and emerges over time, the end points of our intentional mobilization of non-human resources to achieve a goal often have very different outcomes than were initially expected. The divergence results from the agency of the material actants that humans work with. This can be, or rather generally is, the case even when the scientific endeavor is judged to be successful. Pickering stresses the "significant parallels and intertwinings between the intentional structure of human action and material agency. Especially I want to stress the temporal emergence of plans and goals and their transformability in encounters with material agency" (17). He avoids dualism by seeing science and other human endeavors as involving non-humans as partners in a "dance of agency" characterized by the resistance and accommodation of the non-humans being worked with. Material objects, and even more so living non-humans, frequently contribute to the failure of human projects to achieve their desired outcomes, which then require accommodation by humans in response to this material resistance to our intentions. This dance can "include revisions to goals and intentions as well as to the material form of the machine in question and to the human frame of gestures and social relations that surround it" (22). In this way, the differences that they can generate are very significant.

Pickering's focus on machines and non-living matter has limitations from the perspective of scholars like Kohn. Living organisms are selves, in Kohn's term, and have qualities closer to the intentionality of humans than do rocks or jet fuel. For Kohn, material resistance is not the same as agency, which he thinks should be attributed only to life forms. Selves, human and otherwise, are for Kohn (2013: 92) "the product of a specific relational dynamic that involves absence, future, and growth." Absences create actions pointing toward the future fulfillment of needs, such as food or water, and success permits growth, made possible by motivated actions that cannot accurately be described as applying in the same way to non-living matter. We concede the validity of his argument that not all actants are the same. But the acknowledgment by scholars like Pickering of the existence of different kinds of actants, some with intentions, some without, allows us to maintain the analytical strengths of the ANT approach.

These strengths for us lie particularly in the way that ANT provides a method for tracing the entanglements we encounter in conducting holism without boundaries.

The auto-mobility, self-replication, and transformational capacities of the human microbiome fit with Kohn's argument as well as the ANT approach, but the consequences of these features are even more apparent in the case of infectious diseases. Kohn (2013: 94) says that his attempt to open anthropology up to what lies beyond the human involves "finding ways to make general claims about the world. These claims don't necessarily line up with certain situated human viewpoints, like, say, those of animists, or those of biologists, or those of anthropologists." His book is titled, he stresses, "*How Forests Think*, not *How Natives Think, about Forests*" (94). If we limit our understanding of non-human life to what it means for people, and how they perceive it, many important features will be missed. We will argue this point in greater detail in the next chapter, with reference to Clifford Geertz's famous essay on Balinese cockfights.

The belief that a disease was spread by a bad smell (the **miasma theory**, widely held by physicians as well as the general population until late in the nineteenth century) rather than tainted water or food, didn't stop waterborne disease vectors like typhoid and cholera from infecting and killing many people. This is not to say that the belief was not important, because it did influence the kind of health interventions that were undertaken; for example, attempting to live upwind of noxious facilities like stockyards, creating geographical divides that are still visible in many cities (Ghitter and Smart 2008). Regardless of our knowledge or beliefs, disease microbes still make very crucial differences. If they were to have properties other than the ones actually found at present, the consequences could vary as well. For example, if (or, perhaps, when, as a result of rapid evolution among viruses) the current strains of avian flu become transmissible between humans by air, it would be much more difficult to prevent the emergence of a disastrous pandemic.

Science and Explanation in Posthumanism

One risk of this kind of example and argument, helpfully pointed out by a reviewer of the first draft of this book, is that it might be interpreted as an elaborate route through social theory that ends up simply telling us that the natural scientists were right all along, that their approach to understanding nature on its own terms is adequate. All we need to do is to really understand the nature of microbes, and how they can be controlled, for example. Why is posthumanism needed to bring us back to the primacy of natural science's forms of explanation and understanding? And if human scientists are doing the explaining, aren't we simultaneously being returned to the dualistic humanism and emphasis on reason of the Enlightenment?

These questions return us to the debates within studies of science, technology, and society (STS). Space does not allow us to summarize the vast body of work in this field, so we will make only a few key points here. First, despite claims to objectivity, science always incorporates bias. Feminist critics have shown that sexist bias has characterized science since its beginnings, serving for instance to "naturalize" particular scientific constructions of femininity, with ideologies of gender shaping scientific inquiry and practice in distinctive and often surprising ways (Jacobus, Keller, and Shuttleworth 1990). There was a dominant idea in early modern Europe that because of their mothering and birthing, women were closer to nature than men, and modern science tended to operate on the assumption that the male body, under the control of its rational mind, was the standard, with female bodies (and minds) being less fully human (Dumit and Davis-Floyd 1998). Such bias is not only historical: at present, women are seriously underrepresented in clinical trials of drugs, because of complications added to research protocols by menstrual cycles. The results of this bias are immense: "Women are almost twice more likely to have an adverse reaction to a drug than their male counterparts, and 80% of drugs withdrawn from the market are due to unacceptable side effects in women" (Bruinvels et al. 2016).

Linda Fedigan (2001) explored the argument by Donna Haraway (1989) that primatology began to develop along feminist lines in the 1970s at the same time as the second wave of women's movements was cresting. Fedigan asked whether primatology could indeed be seen as a feminist science. Her six criteria, drawing from a wide review of the feminist critique of mainstream science, began with reflexivity, the principle that researchers should pay attention to their own position in the research process, and how this position affects the results that they obtain. The other criteria were "taking the female point of view; cooperation with rather than domination of nature; moving away from dualisms and reductionism; humanitarian applications of science; and greater inclusiveness of formerly marginalized groups" (Fedigan 2001: 47–48). Using this definition, she concluded that primatology had increasingly shown these features over the last quarter of the twentieth century. What she found important was not just that primatology had shifted toward exhibiting the characteristics of a feminist science, but that it had moved to correct the androcentric bias in its practices much more quickly than most other disciplines. However, these changes occurred in the face of many primatologists actively rejecting the idea that feminism should influence how science should be conducted. Instead, the primatologists she interviewed argued that "they changed their practices in order to make their science better," not because feminists thought they should, "but because it was right, scientifically right, to flesh out the picture of female primates, to consider questions from a female, as well as a male, perspective, and to research issues of concern to women as well as men" (Fedigan 2001: 66).

The sociology of scientific knowledge has demonstrated that the interests of scientists and their backers have immense influence on what directions and with which assumptions a field of research develops. A key feature of this approach is that it doesn't distinguish between "true" and "false" scientific accounts, but is concerned only with explaining why those interpretations were adopted (Bloor 1999). While the aims of this scientific project, to see the operation of science in its social and cultural context, is important and has been productive, Fedigan's (2001) analysis of primatology suggests that the concern of scientists with getting at the "truth" can, under certain circumstances at least, impel them to develop a field of research in certain directions that may not necessarily suit their "interests," other than those of getting it "scientifically right."

Bruno Latour (1999) argued that the sociology of the approach to scientific knowledge, with its distinction between "nature" and "social interpretations of nature," created a situation where non-human actants could make no difference, dropping out of the story, which focused on which interpretations of phenomena were successful in being accepted. The approach, he argued, in essence reproduced the modernist dualism where all the motive force in change was assumed to come from people. Yet, if we take "material agency seriously, on its own terms ... we yield up our analytic authority to the scientists themselves. Scientists, not sociologists, have the instruments and conceptual apparatus required to tell us what material agency really is" (Pickering 1995: 12). At best, social scientists become popularizers and translators of what the "real experts" have learned.

Our first response to this quandary is to suggest that if we are to avoid anthropocentrism in the era of the Anthropocene, when our collective decisions have global consequences, we cannot afford to reject solid scientific evidence about the non-humans with which we share the world. We cannot rely on our cultural assumptions about them but need rigorous information about how forests "think," how they develop their own emergent dynamics, which affects the ecosystem in ways that we are only beginning to understand. Postmodern deconstruction of the underlying cultural foundations of scientific accounts is not enough. However, deconstruction and critique are still important, because natural science does have biases, interests promoted by powerful governments and corporations, unfounded assumptions, and tunnel vision that leads it to neglect unconventional approaches and ideas. Our experience with the neglect of non-human life in mainstream border studies is a clear illustration of the problems raised by anthropocentrism in a field of research.

Second, the science that we need and can rely on for guidance in the singular epoch we now find ourselves in, is not the science of a few decades ago. What we need is not a science that is built on a dualism between nature and society and a complete neglect of the insights of the social sciences and humanities. If the world is a tangled web of radically different kinds of actants interacting with complex and unpredictable consequences, narrow specialization is not enough for our needs, however necessary it might be to attain the rigor of traditional scientific knowledge. Fortunately, even as many social scientists have accepted that they have much to learn from the natural sciences, the same movement is happening on the other side, with biologists such as Maturana and Varela (1972) giving a great deal of attention and thought to social theory. The exciting possibilities of serious and generous convergence and collaboration can be seen vividly elaborated in Tim Ingold's (2000) book *The Perception of the Environment: Essays on Livelihood, Dwelling and Skill.*

Third, while we may seem to come back to the conventional insights of the natural sciences, the journey through the sociology of scientific knowledge, ANT, and Andrew Pickering's work brings us not just full circle, but beyond the starting point to a newer appreciation of what actually takes place in the doing of science. Material agency, whether of parasites or the oil sands, needs to be seen as itself entangled with the human world and the activities of the scientists. New epidemics result from past medical breakthroughs, as bacteria become resistant to antibiotics, for example. To deal effectively with the threat of antibiotic resistance requires not just new drugs or even new ways to manipulate the microbiome to build up its defenses. It also demands new ways of regulating their use; educating the public about the problems not only of overuse but also of not completing their full prescribed course of usage; and finding ways to stop doctors from prescribing antibiotics as a quick way of convincing patients that they are actually doing something, even if the drugs are completely ineffective against viral infections.

Finally, Andrew Pickering's work brings back symmetry between humans and non-humans in science in a useful new way by showing that the distinction between the knowing subject (the scientist) and the material objects on which they work is only part of the process. He argues that science is only partially about finding ways to represent in scientific language the operation of matter and organisms. Instead, he says,

> My basic image of science is a performative one, in which the performances—the doings—of human and material agency come to the fore. Scientists are human agents in a field of material agency which they struggle to capture in machines. Further, human and material agency are reciprocally and emergently intertwined in this struggle. Their contours emerge in the temporality of practice and are definitional of and sustain one another. (Pickering 1995: 21)

In other words, both material actants and scientists do things. Scientists are not "disembodied intellects making knowledge in a field of facts and observations" (Pickering 1995: 6), but instead struggle with machines, matter, and organisms to get their experiments and projects to work. Pickering's way out of the dualism of culture and nature is to pay more attention to what human and material agency have in common: "doing things, things consequential in the world. Performance is the place we should start from, not reason, language, representations, symbols" (Pickering 2013: 25–26). Science and technology are practices in which human agents try to deal with the resistance of material agency. Certain ways of dealing with the matter at hand work; others don't. But even when they work, that does not necessarily confirm that the representational system that led to the successful practices is correct. Draining swamps to get rid of miasmatic causes of disease often did reduce the prevalence of diseases transmitted by infected water sources.

Microorganisms and Borders

History clearly demonstrates the consequences of microorganisms; that is, the effects of their material agency on humans, which we had no knowledge of before the nineteenth century. Until the fifteenth century, the world's two hemispheres were almost completely isolated. One of the results of the Age of Exploration was that diseases and parasites were transmitted from one part of the world to another that had developed no immunities. This **Columbian exchange** of diseases between New and Old Worlds brought devastation to populations without immunities. It has been estimated that as much as 80 to 95 per cent of the Native American population was decimated as a result of introduced diseases, exacerbated by other mistreatments such as enslavement that heightened their vulnerability, within the first 100 to 150 years after 1492 (Nunn and Qian 2010).

As we mentioned in Chapter 1, we first became convinced of the utility of posthumanist ideas while thinking about the movement of non-human life across borders. Microbes had an impact on border crossing long before we had any idea of their existence. The first passports that are known in history appear to be individual health passports that the Florentine Board of Health established during the 1348 outbreak of the plague (Porter 1999: 36). Entry to the city required these bills of clean health. Passports in the modern sense didn't appear until almost two centuries later. After war with the Ottomans ended in 1739, the Austro-Hungarian Empire created a large plague-control zone in Croatia that employed 4,000 troops. Sentry posts and patrols had orders to shoot unauthorized travelers. People coming from Ottoman lands had to submit to groin and armpit strip searches and a quarantine as long as 48 days. Trade goods were fumigated. Suspect raw wool was put in a "warehouse where low-status people were made to sleep; if they developed plague symptoms they were shot and the wool was burned" (Watts 1997: 25). In the absence of antibiotics and accurate knowledge of disease transmission, quarantines were the most practical way of dealing with epidemics. The tendency of people to flee from affected areas undermined the effectiveness of quarantines. The United States and Canada had open borders for migrants until the beginning of World War I, with several exceptions, including the Chinese Exclusion Acts and for political radicals (Smart and Smart 2012a). The most important limit was mass medical screening upon entry at Ellis and Angel Islands in the United States and at Pier 21 in Canada.

As we move about in the contemporary world, we bring with us both the beneficial elements of the microbiome and any predatory infectious diseases and parasites we may carry. Immense public health risks result from the combination of this situation and increased mobility, but management strategies are limited by inadequate surveillance and reporting systems. Since the 9/11 attacks on the World Trade Center and the postal anthrax attacks of 2001, concern for biosecurity has grown rapidly. Talk about biosecurity brings together diverse issues and actors concerned that new biological threats challenge existing approaches to security (Collier and Lakoff 2008: 8). Fidler (2010: 290) notes convergence of policies on biological weapons and infectious diseases because those concerned with biological weapons realize that public health has "become critical to their mission"; framing infectious diseases as threats to security increases the political significance of public health. The term "biosecurity" entered the legal lexicon only in 1993 with the enactment of New Zealand's Biosecurity Act, but "the *idea* of biosecurity stretches back to pre-colonial New Zealand"(Dunworth 2009: 157). New Zealand's biosecurity agency, however, emphasizes protection from the risks posed by pests and diseases to the economy, environment, and people's health, whereas since 9/11, attention around the world has been turned to bioterrorism and the risk of pandemics.

Pandemic crises have made it clear that surveillance was previously weak. No public health warnings had been disseminated to hospital authorities in Toronto before the first SARS cases there in 2003, despite reports that had been appearing for months before in Hong Kong newspapers about a new form of atypical pneumonia (Smart and Smart 2008). Surveillance by national and international public health agencies has since expanded. The influence of the World Health Organization (WHO) has grown over the last decade. Fidler (2010: 298) concluded that the WHO's 2005 International Health Regulations impose obligations on states "never before seen in international law on public health." Whether the biosecurity emphasis on pandemics is the most cost-effective expenditure of scarce public health resources, however, is questionable. In contrast to those who see newly emerging diseases as a result of technological change (travel plus increased use of antibiotics, etc.), some instead point to the collapse of basic public health systems. From this perspective "global living conditions—poverty, civil war, lack of basic healthcare—were the source of the emerging disease threat," and these social problems must be addressed to "provide security against emerging pathogens" (Lakoff 2008: 47). As Paul Farmer (2001) has stressed, the plagues of the poor cause much more misery but receive much less funding. The need for a holistic perspective is clearly displayed in these discussions: a narrow medical focus on disease pathogens cannot succeed, because the problem emerges from complex constellations of very disparate influences, from transport technology to inequality and governmental approaches that neglect basic public health measures. As Alex Nading (2014: 10) argues in a study of dengue fever control in Nicaragua, where neighborhood teams search homes for neglected surfaces that can allow mosquitoes to breed, entanglement with disease organisms means that we cannot simply assume that the global affects the local; instead, "changes in bodies reverberate through landscapes, and vice versa ... dengue renders the scalar distinction between local and global infrastructures, bodies, and forms of knowledge increasingly difficult to maintain. Dengue makes the ostensibly intimate operations of home life a public concern, and it drives public concerns into the center of intimate life" (Nading 2014: 10).

Zoonoses

Zoonotic diseases in particular have become central to the biosecurity concerns of governments in the last decade. It has become clear that this class of diseases is not exceptional, but rather almost certainly the main source of new human diseases. Infectious disease is "all around us. Infectious disease is a kind of natural mortar binding one creature to another, one species to another, within the elaborate biophysical edifices we call ecosystems" (Quammen 2012: 20). About "60 per cent of all known human infectious diseases either routinely cross or have recently crossed between other animals and humans" (20). Many of the rest crossed the species barrier earlier.

David Quammen (2012) points out that the rapid growth of zoonotic emerging infectious disease, including SARS, Ebola, and Human Immunodeficiency Virus (HIV), reflects

> the convergence of two forms of crisis on our planet. The first crisis is ecological, the second is medical. As the two intersect, their joint consequences appear as a pattern of weird and terrible new diseases.... How do such diseases leap from nonhuman animals into people, and why do they seem to be leaping more frequently in recent years? To put the matter in its starkest form: Human-caused ecological pressures and disruptions are bringing animal pathogens ever more into contact

with human populations, while human technology and behavior are spreading those pathogens ever more widely and quickly. (39)

The Anthropocene, and all of the human activity that has brought us to it, fosters zoonotic diseases, as we displace and disturb the natural habitats and hosts of diseases that once plagued only other animals. These new epidemics display Bruno Latour's hybrids in profusion and show why the production of hybrids across the nature–society divide is vitally important. The last part of this chapter offers BSE as an extended example to demonstrate how actor-networks operated to produce the crisis and how responses to the crisis have changed the world in some fundamental ways, bringing into being networks that span the scales from microscopic to global.

ANT has been summarized in a single phrase: "Just observe and describe controversies" (Venturini 2010: 2). Controversies are "situations where actors disagree"; they "begin when actors discover that they cannot ignore each other and ... end when actors manage to work out a solid compromise to live together" (4). From a different perspective in science studies, Harry Collins and Trevor Pinch (1998: xv) have argued that for citizens who want to participate in the democratic processes of technological societies, "all the science they need to know is controversial." Their Golem series provides readable introductions to a variety of science that is, or was, controversial, shedding clear light on how observations and data can be interpreted in dramatically different ways and with significant consequences for the development of the field of research, and often society more generally.

ANT is posthumanist and non-anthropocentric in a stronger way than simply accepting non-humans as actants, as having agency. Scholars using the approach do not see "actors" as something permanent and solid. Instead the "actors" in a network are seen as things that are brought into being by networks, and these networks help to determine which things serve as actors and actants. During a controversy, "any actor can decompose in a loose network and any network, no matter how heterogeneous, can coagulate to function as an actor" (Venturini 2010: 4–5). Time and the unfolding of interactions in historic contexts are thus crucial to any specific ANT analyses. The "BSE crisis" will be shown to operate this way. Once scientifically accepted and incorporated into the rules of national and international agencies, just a reference to this crisis could enable vast actions, as we will see below. Actants, remember, are simply anything that makes a difference in an association. Humans, particularly the roles that we play in a particular situation, can be decomposed into the product of a variety of networks, while scallops or prions can be actants within networks. The nature and causes of BSE, as well as what should be done about it, are a classic example of such a contested situation.

Controversies often revolve around newly described entities such as microbes, X-rays, or prions. Latour presents scientists and engineers as "mobilising large numbers of allies, evaluating their relative strength, reversing the balance of forces, trying out weak and strong associations, tying together facts and mechanisms" (Latour 1987: 240). Science in the making, as opposed to the accepted facts, entities, and phenomena of stabilized science and technology, "normal science," as Thomas Kuhn (2012) described it, involves controversies that can be resolved only by scientists and their allies mobilizing and deploying resources more effectively than their rivals (Latour 1987). What we call "scientific facts" reflect compromises that emerge gradually in the aftermath of controversies. Past uncertainties are forgotten and left to historians, at least until a new controversy renders them once more subject to challenge. Once settled, they serve as "black boxes": their properties, particularly knowledge about what inputs produce what outputs, can simply be accepted and incorporated into high-level discussions, at least until they become questioned once again. Black boxes are everywhere in our daily life; for example, a door handle that operates a latch. You learn as a child what you have to do to make it open, while having little, if any, understanding of the mechanism inside. Andrew Pickering (2010) uses this idea of black boxes to explain his ideas of science as involving much more performance than representation, discussed earlier in this chapter. A black box is something that we do something to in the dance of agency he emphasizes. Even in the practice of science, innumerable black boxes are used, and knowledge of their workings is not a prerequisite for working with them. Such knowledge is something that may or may not develop out of our performative experience of the box. The conventional view of science, he says, is as an unquenchable thirst to open and understand all black boxes. Each scientific research team, though, is generally trying to open only one black box at a time, while using a vast number of others as tools to get on with the work. So at least in terms of what is most common in mundane scientific practice, we would conclude that using black boxes is more prominent, if less emphasized, than trying to open them and understand them. In the efforts to open up the black box of BSE, discussed below, innumerable other black boxes were being relied on to do this investigative work, and at various times those subsidiary black boxes themselves became controversial.

BSE is now known to be part of a family of diseases called transmissible spongiform encephalopathies (TSEs). Prior to BSE, the best known member of this family was a disease of sheep known as scrapie, the cause of which was a mystery, since no bacterium or virus could be identified. The term "prion," or "proteinaceous infectious particle," was coined by Stanley Prusiner (1997) to distinguish it from viruses. Prusiner received the 1997 Nobel Prize in Physiology or Medicine for isolating the scrapie disease agent. His theory, suggesting that infectious proteins transmitted this kind of disease, was very controversial, to the extent that Prusiner described himself as a "heretic," because a protein does not contain the DNA needed to allow it to reproduce by infecting other cells. It is thus not a life form in the same sense as bacteria, viruses, or other microbes are. Arguably, it is not living at all, despite its capacity to convert other protein molecules into toxic forms, and thus poses a question for Eduardo Kohn's argument that living organisms are fundamentally different from non-living matter in ways that make it appropriate to refer only to life forms as agents.

Prusiner's theory was neither the first controversial explanation of TSEs nor the first Nobel Prize award received for work in the field. Carleton Gajdusek had that distinction in 1976 for his work on **kuru** in Papua New Guinea. Kuru was a disease involving tremors, rapid deterioration, and death, first seen among the Fore of Papua New Guinea in the late 1920s. Government responses began only in 1953. The first medical evaluation in 1955 suggested that it was a case of "acute hysteria" thought to be "precipitated by the threat or fear of sorcery." This explanation was rejected by Gajdusek when he observed anatomic evidence of advanced neurological decay (Lindenbaum 2001: 368).

Shirley Lindenbaum and Robert Glasse began their anthropological study of kuru in 1961. The research was funded by the Rockefeller Foundation through Henry Bennett at the University of Adelaide. Bennett asked them to document Fore genealogies, their patterns of descent from ancestors, one of the longest-standing research techniques in anthropology, dating from before fieldwork began in the nineteenth century. It soon became apparent to them that many of the kuru victims "were not closely related biologically, but were kin in a non-biological sense" through cultural kinship classifications (Lindenbaum 2008: 3715). Kuru had spread slowly through Fore villages within living memory. Its "progress through Fore territory followed a specific, traceable route," a finding at odds with a purely genetic explanation (3716).

They found a relationship between kuru and cannibalism. Although cannibalism was no longer practiced in the 1960s, after its suppression by the government and missions, "the Fore spoke openly about the recent customary practices of consuming the dead" (Lindenbaum 2008: 3717). Cannibalism as the source of transmission for kuru gains support from their finding that other than the Fore, it was customary for groups in the region to "consume enemies (exocannibalism), not deceased kin (endocannibalism) ... a pattern of behaviour with consequences for the transmission and geographical boundaries of kuru" (3717). All body parts were eaten, "except the gall bladder that was considered too bitter. Not all bodies were eaten. The Fore did not eat those who died of dysentery, leprosy and possibly yaws, but kuru victims were viewed favourably" (3717).

Their findings received "little, often sceptical attention, until the anthropological and medical stories came together in 1966, when chimpanzees injected with brain material from victims of the disease exhibited a clinical syndrome akin to kuru" (3718). This finding gave "credence to the cannibalism hypothesis, as did the fact that following a change in this mortuary custom, kuru disappeared among children, while the age of those afflicted with the disease also rose" (3718).

Gajdusek's success at identifying the source of kuru made him famous. He became director of two National Institutes of Health programs in the United States and was described as being at the hub of "an empire of perhaps two hundred collaborating laboratories worldwide" (Roger Bingham, quoted in Rampton and Stauber 2004: 54). He hypothesized that kuru and scrapie were caused by "slow viruses," which caused slowly progressing symptoms, unlike the rapid infections of classic viral diseases such as smallpox or measles (Rampton and Stauber 2004: 49). This general idea proved fruitful, and a variety of viruses were found with long incubation times, among them HIV. However, the TSEs were not among them. This case demonstrates the need for proponents of a scientific explanation to enroll diverse networks of allies and physical support (well-preserved brains for examination in well-equipped laboratories with well-trained researchers) in order for their explanation to attain the status of "scientific fact." When sufficient allies and resources are mobilized, new scientific claims can move rapidly to cover, and sometimes fundamentally change, the globe. But the case of kuru also shows that the black boxes of established explanations, even those proposed by scientists who become Nobel laureates for their work, are always subject to new controversies that may bring them into question. That is what happened with Prusiner's alternative explanation of TSEs by prions, which was hotly contested at the time, since proteins proliferating by themselves without the aid of DNA and RNA ran counter to standard ideas about disease causation. Skepticism about the prion as cause of BSE persisted among several farmers that we talked to during our research in Alberta (2006–10). One prefaced his agreement to be interviewed about the impact of BSE by warning us that he and his wife didn't believe in BSE. Instead, they were proponents of the alternative theory popularized by Mark Purdey, a British farmer, who argued that the cause was the use of the chemical phosmet to treat warble fly infection. Alternative hypotheses about BSE "reflect anxieties about the environment, industry, the government, and the intrusion of outsiders" (Lindenbaum 2001: 379).

Mobilization of allies is a central aspect of the response to the discovery of BSE in North America, in forms ranging from approved BSE testing regimes, import bans, modified risk classifications, World Trade Organization (WTO) rulings, risk evaluations based on the precautionary principle, recruitment of public support, and court rulings. Scientific consensus now agrees with the basic story about BSE. In the brains of cattle, a protein misfolded and developed very different properties: it converted other proteins around it into the misfolded kind, and then it spread. These spreading abnormal proteins eventually formed plaques and left spongy holes in the brain: the "spongiform" of the name. The damage to the brain resulted in tremors, loss of motor coordination, the inability to walk, and then death. Originally thought to be a disease only of cattle, in 1996, the UK government confirmed growing beliefs that it could be transmitted to humans who ate tainted beef. This announcement set off a global crisis.

Up to 482,000 BSE-infected animals had entered the human food chain before controls on specified risk materials (brain, spinal cords, lymph nodes, etc.) were introduced in 1989. Millions of consumers would have eaten beef from infected cattle. Humanity was very lucky with BSE, with a cumulative total of 226 cases of variant Creutzfeldt-Jakob disease (vCJD) in humans (all fatal), again because of the nature of the prion. First, it turned out that the risk from eating most cuts of beef was minimal. Second, it appears that few of those exposed to specified risk materials developed vCJD, and that there is only a small proportion of the population with a particular gene variant that makes them susceptible to the infection.

From an ANT perspective, what was particularly important about this new entity, the prion, was that it was not destroyed by conventional rendering heat treatments. BSE first spread among cattle through the practice of rendering cattle parts and deadstock into meat and bone meal, which was often fed to cattle as well as to other farm animals. If the prion hadn't had those unusual properties, the number of cases would have been very low, based on a very low rate of spontaneous mutation. Without human intervention, the prion would not have been recirculated into the cattle population. The human health dimension to the BSE outbreak would have been even smaller. It would have been an animal health problem, with at most very small numbers of humans affected, comparable to foot and mouth disease. Questionable agricultural practices, which were widely criticized as involving the "unnatural" act of turning vegetarian cows into cannibals through the ingestion of bovine meat and bone meal, were the ground on which the zoonotic crisis grew (Lindenbaum 2001). ANT tells us that conditions influence when and where an entity will make a difference in a network of events. Traditional livestock raising practices would not have "amplified" (through feeding cattle carcasses to cattle) the BSE that developed spontaneously or due to genetic predispositions (at very low rates). But the properties of the prion itself were crucial in making a difference to the networks of agricultural production and food distribution and consumption.

Most experts agree that BSE was a "historical watershed with regard to the way food is thought about, talked about, and handled" (Lien 2004: 3). BSE in the UK and elsewhere in Europe was a catalyst for widespread changes in food safety and agricultural regulation. In the context of global trade, "even countries that had no strong domestic reason to improve their **food safety**, had an international imperative to do so—if they wanted to gain and maintain access to international markets" (Spriggs and Isaac 2001: v). Food politics became more contested due to the increased distance between producer and consumer, along with the decline of public trust in expert authority (Beck 1992; Lien 2004). Erosion of trust in regulatory agencies and scientific experts is particularly marked in Europe, whereas faith in technology and industry, if not government, is much stronger in the United States. Canadians have more faith in their governments than Americans (Skogstad 2006: 221).

When the first indigenous Canadian BSE case was confirmed on May 20, 2003, world reaction in the form of export bans was immediate. For each day of the trade bans, the Canadian economy lost an estimated $11 million and many jobs in multiple sectors linked to the cattle industry (Le Roy and Klein 2005). Canadian BSE can be traced to the 1990s importation from the United Kingdom and Ireland of animals and animal feed derived in part from rendered cattle. Between 1982 and 1990, 191 cattle were imported. It was alleged in a BSE class action suit that as many as 80 of these infected cattle may have been rendered into the Canadian cattle feed system. Importation of cattle from the UK stopped in 1990, but importation of UK meat and bone meal cattle feed was not banned until 1997. The most likely source of the first domestic case was feed produced in St. Paul, Alberta, in 1997, a few months before a mammal-to-ruminant feed ban took force. A class action suit was launched against the Government of Canada and Ridley Inc., the producer of the feed. The suit alleges, among other faults, gross negligence on the part of the Government of Canada.

Canadian beef was particularly vulnerable to disruptions deriving from export restrictions because of its heavy reliance on exports of live animals, up to 60 per cent of total production in 2002. A patriotic response from Canadian consumers who wanted to support beleaguered farmers made Canada the first country in which beef consumption increased (slightly) rather than dropping sharply after the discovery of BSE (Smart 2008). This local consumer support was in part due to framing the issue as another case where the US was engaging in unfair trade politics rather than a "real" food safety problem. Increased domestic demand could not resolve the problem, though, because domestic slaughter capacity was insufficient to cope with the cattle population, particularly for cull (older) cows, normally used for hamburger.

The federal and some provincial governments spent over $2 billion on recovery programs intended to help producers and the cattle industry cope with the aftermath of the 2003 BSE outbreak. The effects of the recovery programs, though, were more complex than the simple provision of a safety net. Some parts of the beef industry benefited more than others. The recovery programs were seen by many producers as poorly designed and almost exclusively helping the large feedlot operators and American-owned packing plants. Alberta's auditor general concluded that most of the support funds benefited large firms, mostly American controlled, but pointed to poor program design rather than fraud (Dunn 2004). The kinds of networks that small and medium producers had were very different from those of the corporations that ran the meatpacking plants, so that the corporations could more easily incorporate the prion into advantageously restructuring their networks. The different parts of the beef-production chain were affected by prions in diverse ways. The situation seems to have contributed to the decline of small, independent producers. Their displacement might ultimately erode the imagery and reputation of the industry that propped up consumer confidence.

Two key elements of Canada's response to BSE have been the enhanced feed ban, in force since July 1, 2007, and the segregation during the slaughter of under-30-month (generally considered free of BSE transmission risk) and over-30-month cattle. The policy responses have been of great importance in ending cattle-to-cattle transmission through feed. However, the requirement to set up two separate production lines in slaughter facilities has been associated with substantial increases in slaughter costs that have made it very difficult to create or sustain smaller abattoirs. Various conditions in the enhanced feed ban have undermined the viability of pre-existing provincially inspected slaughter facilities, and made it more difficult for producers to market premium-niche products such as organic, "natural," or grass-finished beef. An unintended consequence of the feed ban is that some farm families are returning to slaughtering their own animals for subsistence use. Another example is that since it now costs $75 to have a deceased animal picked up for rendering, many producers simply dump their losses in the back quarter to be scavenged by wildlife such as coyotes and crows. In these ways, a policy intended to increase control may actually result in less control over certain aspects of the handling of slaughter and carcass disposal.

On the global scale, BSE is one of a series of food-related crises and conflicts that have heightened conflicts over food and agricultural trade: BSE, SARS, avian flu, swine flu, genetically modified crops, growth hormones in livestock, tainted pet food and cough syrup, and so on. Food and agricultural products raise concerns different from those for mechanical products because of food's inherent association with organic materials that can spread infectious diseases (Smart and Smart 2012b). Stricter standards, like all policy interventions, have a variety of unintended consequences. Particularly given the high cost of participation in the setting of international standards, magnified by the complicated and contested nature of global agri-food governance institutions (Busch 2004), standards are at great risk of being developed in ways that marginalize smaller firms and smaller countries. They may ultimately serve the long-term interests of the largest agri-businesses, even if in the short term these companies argue for less, rather than more, state interventions (Freidberg 2004: 220).

Our extended example of BSE shows how microscopic entities were constructed as "scientific facts" and became actants that are still making a difference, jumping from the microscopic to the national and global scales through the kinds of networks they allow to form, and the differences they make in the action that occurs. Hybridization between humans and non-humans due to modern technology and science changed the world in the case of BSE, as well as in many others. The resultant changes worked out in ways that were neither expected nor intended. Given that this could have developed in different directions if other circumstances had been in place, the case seems to provide an excellent illustration of what Pickering calls the dance between human agency and material agency, where the nature of the material partner structures the outcomes through its resistance (e.g., to rendering at conventionally high temperatures) and human accommodations to this resistance (e.g., by specifying certain high-risk parts of the cattle body to be excluded from the food chain of both humans and cattle). But when the dancing partners are microbes, we can no longer assume that it is the human partner who takes the lead.

In the next chapter, we shift our discussion to human relationships with non-humans by looking outside our body at creatures visible to the naked eye. The dances of agency that they engage in with us are both more visible and more subject to strong relations of connection and solidarity between non-human animals and ourselves. How these relationships are understood and how they affect social organization has had great consequences throughout our history, but are changing once again in interesting ways.

Discussion and Activities

Look at the places where you store medicines, pharmaceuticals, and household cleansers. How many different defenses do you possess against microorganisms? Why do you possess and use them? Discuss whether there are other antibacterials in your household that you don't normally think about.

Find a current controversy relevant to topics addressed in this chapter, through news or social media. How do different positions about this controversy link together different humans and non-humans in networks? Can you describe a "dance of agency" between the intentions of humans and the resistance of the non-humans that are involved in the controversy?

Google search a recent infectious disease scare (Zika is current as we write this, for example). What are the issues that are controversial? Do different interpretations of the problem involve proposals for different relationships between people and non-humans (e.g., canceling the Rio 2016 Olympics due to Zika concerns)?

Additional Readings and Films

Readings

Dunn, Rob R. 2011. *The Wild Life of Our Bodies: Predators, Parasites, and Partners that Shape Who We Are Today.* New York: Harper.

Farmer, Paul. 2001. *Infections and Inequalities: The Modern Plagues.* Berkeley: University of California Press.

Lindenbaum, S. 2001. "Kuru, Prions, and Human Affairs: Thinking about Epidemics." *Annual Review of Anthropology* 30: 363–85.

Quammen, David. 2012. *Spillover: Animal Infections and the Next Human Pandemic.* New York: W.W. Norton & Company.

Watts, Sheldon. 1997. *Epidemics and History: Disease, Power and Imperialism.* New York: Yale University Press.

Zimmer, Carl. 2012. *A Planet of Viruses.* Chicago: University of Chicago Press.

Films

Contagion. 1987. Film directed by Karl Zwicky. Australia: Premiere Film Marketing Ltd.

Contagion. 2011. Film directed by Steven Soderbergh. Beverly Hills: Participant Media.

The Masque of the Red Death. 1964. Film directed by Roger Corman. Los Angeles: Alta Vista Productions.

The Masque of the Red Death. 1989. Film directed by Larry Brand. Los Angeles: Concorde Productions.

The Plague of Florence. 1919. Film directed by Otto Rippert. Germany: Decla Film-Gesellschaft.

The Omega Man. 1971. Film directed by Boris Sagal. Los Angeles: Walter Seltzer Productions.

MULTISPECIES ETHNOGRAPHY

The central claim of this book is that to better understand humanity, the primary subject of anthropology, we have to enlarge our focus beyond humans. For anthropologists, this requires attending to non-humans in our research. Although not the only research approach used by anthropologists (e.g., direct observation by primatologists, interpreting prehistoric material culture for archaeologists), ethnography is our focus for this chapter. While ethnography is sometimes used in other disciplines to refer to non-standardized interviewing as the only method for data collection, anthropologists expect to use multiple methods to collect information and pursue insights, including participant observation and "being there" for substantial periods. But method is only one part of what anthropologists usually mean by "ethnography." The other meaning of ethnography is the detailed description of a human way of life, a documentary effort that has been central to social and cultural anthropology for over a century. In the last chapter, we considered people as thoroughly mixed up with microbes. The human body is a combination of human and non-human components, and our world is profoundly influenced by these mixtures. In this chapter, we consider our relationships with animals, but with some attention to plants as well. Throughout human evolution, and in every place on earth, even the most urban and "artificial," we have not only been surrounded by animals, but have been changed through our interactions with them, as have they. Indeed, animals have been "central to the formation of every human society and mode of production on the planet" (Coulter 2016: 5). As shown in Chapter 2, with the world-changing Colombian Exchange, both New- and Old-World ecologies were transformed by both introduced and invasive species that came with the new trade and conquest linkages.

After a discussion of the consequences of allowing human–animal duality to determine the scope of anthropology by excluding non-humans from our focus, we examine several efforts to go beyond such limitations. Then we consider the influential approaches of interpretive and postmodern anthropology, suggesting that, despite their merits, they suffer from the limited and exclusivist scope of humanism. Explaining this will require a brief excursion through the development of the discipline of anthropology. Finally, we discuss some examples of ethnographic work that includes other species, suggesting the ways in which a broader and more inclusive holism enhances not only our knowledge of them, but also of us.

Human Exceptionalism

Posthumanism challenges dominant ideas of "ourselves" as fundamentally different from other forms of life. This **human exceptionalism** blinds us to many processes that involve multispecies interactions (Tsing 2012). Many perspectives rely on dualistic divisions between nature and society, or nature and culture, both implying a binary distinction between human and nature. Following from this kind of worldview, by which we mean a broad understanding of the nature of the world and universe in which we live, to say what the word "human" means is almost always to distinguish it against "animal." Even when we acknowledge that we are also animals, most people feel the need to stress that, in some ways, usually thought to be the most important ways, we are much more than animals (DeMello 2012). We are "tool-using animals," "tool-making animals," "thinking animals," "language-using animals," and so on. Philosophers have rarely asked, What makes humans animals of a particular kind? The much more common question is, What makes humans different in kind from animals? The relation between the human and the animal is thereby "turned from the inclusive (a province within a kingdom) to the exclusive (one state of being rather than another)" (Ingold 1994: 19).

If we follow Andrew Pickering's path of stressing greater commonalities between material and human agencies when we focus on performance (doing) rather than representation (knowing), this kind of dualism starts to erode. An excellent example of such convergence can be seen in Olga Solomon's (2012) study of cross-species sociality. Her research highlights the importance of being open to surprise, a central feature of the ethnographic imagination that is not always present in other methods that emphasize the testing of hypotheses. At a dog park one day, her border collie dropped a Frisbee at the feet of a five-year-old girl and assumed the play posture. Immediately, the girl began to play with the dog. A commonplace event, except that the girl had autism, and her father was overjoyed, saying that his daughter had never played with anyone before. Since autism has often been thought to be the "antithesis of sociality," this encounter sparked her research on the interaction between children with autism and therapy dogs. Her research undermines a basic assumption about sociality, namely, that sociality is tied up with language, and involves mental states of intention that are characteristically human. She concludes that sociality "is almost never about *being* social, but is almost always about *doing* something together, and *becoming* different in the process" (Solomon 2012: 122). Seen as a performance, rather than a representation, it becomes much easier to understand cross-species sociality, and explain its powerful benefits in therapeutic, and other, contexts.

Many of our most fundamental ideas about humanity are bound up with ideas about (other) animals (Calarco 2015; Wolfe 2010). Religions usually have central beliefs that revolve around the differences and/or similarities between humans and non-humans, whether they are spirit guardians or our dominion given us to steward by God. The monotheistic religions emphasize human mastery of nature (Tsing 2012: 144). The humanity–animality dichotomy frames understanding and treatment of non-humans around us. Humanist ideals urge us to transcend the bestial elements of human nature, and cultivate our purer and enlightened dimensions. The condition of being human is often thought to be "achieved by escaping or repressing not just its animal origins in nature, the biological, and the evolutionary, but more generally by transcending the bonds of materiality and embodiment altogether" (Wolfe 2010: xv).

The boundary between human and animal is not stable: the history of slavery, colonialism, and genocide shows that the category of "animals" is not simply based on zoological classification but is also a political resource that can strip rights, privileges, and even life itself from members of *Homo sapiens* (Braidotti 2013; Wolfe 2012: 10). Cary Wolfe uses this instability of the categories of human and animal to ask what it would mean to grant animals rights as persons, not just as creatures deserving protection. Just as eighteenth-century humanism was racist and sexist, is humanism now **speciesist** (Wolfe 2012)?

For Tim Ingold (1994), Western ideas of humanity have been based on views of animality as a "deficiency in everything that we humans are uniquely supposed to have." Yet we are also reminded by researchers that "human beings are animals too, and that it is by comparison with other animals that we can best reach an understanding of ourselves" (14). In being opposed to humanity, animality suggests life in the state of nature, impelled by instinct and need rather than rational thought. This opposition becomes aligned with those between "nature and culture, body and mind, emotion and reason, instinct and art, and so on" (21). Dualism is also reflected in the division of the study of human nature between the natural sciences, social sciences, and humanities. Humanists encouraged people to strive to escape their baser instincts and to pursue their higher potential through ideals, reason, and scientific knowledge.

There is a paradox at the heart of Western thought: it insists both that humans are animals, in biological terms, and that animality is the opposite of humanity. One response to this paradox argues that "humanity" refers to a biological category (*Homo sapiens*) while being human "refers to a moral condition (personhood)." The underlying assumption is that only members of the human species can be persons (Ingold 1994: 23). This position is currently under legal, political, and cultural challenge from animal rights activists and others. It is also a position absent from many societies, particularly those that traditionally practiced livelihoods based on hunting and gathering (also known as foraging); these communities commonly see animals as also being persons (Descola 2013; de Castro 1998).

Divisions between natural science and cultural explanations of human interactions with animals are not helpful because they take for granted precisely the "separation, of the naturally real from the culturally imagined, that needs to be put into question" to understand people's own perceptions of the world (Ingold 2000: 9). If anthropologists hope to account for the variety of ways in which the world's peoples perceive and understand the world and our companion species in it, we cannot begin from a division between the natural sciences and the social sciences, but need to reinvent both for the task of reimagining our interactions and collaboration in the era of the Anthropocene (Tsing 2015). The traditional division of labor is that scientists tell us how the natural world works, and humanists (including most cultural and social anthropologists) tell us how people interpret that world. This position is consistent with the usual approach to **cultural relativism**, a principle central to anthropology.

From our perspective, cultural relativism is most useful when it is seen not as an ethical position (i.e., that what is good or bad is relative to a particular culture's values) but as a productive way to do research among people with very different worldviews and ways of life, including both foragers and bankers. Cultural relativism tells us that ideas and behaviors are best seen in relation to the social context in which they take place, and thus is closely related to classic holism. Interpreting ideas and behaviors through outside ideas and worldviews introduces distortions and misunderstandings. If we unquestioningly impose our own culture's interpretation on another culture—that is, if we are ethnocentric—we are likely to misunderstand why they do the things they do. This does not carry the ethical implication that we should accept the morality of what they do, making it possible for anthropologists to apply a cultural relativist research approach to people we fundamentally disagree with, such as white supremacists.

Using cultural relativism as a research strategy may be useful, but it does not fundamentally challenge understandings of the natural world. Nature is generally assumed to be the same for all, just understood differently. This assumption can be described as multiculturalism. It accepts that the world is seen and represented by different people in different ways, but presumes that their interpretations of the world are developed in the context of a common shared reality. Some anthropologists have argued that this does not go far enough. To understand the whole range of human ways of life, they tell us, we need to go beyond multiculturalism and proceed from the position that local actors live and experience different worlds, rather than simply representing differing interpretations of a common world.

Multinaturalism

Eduardo Viveiros de Castro (1998), an anthropologist specializing in the Amazon, says that Western ideas of multiculturalism are based on assumptions of the "unity of nature and the plurality of cultures"—that is, while cultural interpretations of nature, and other things, are very diverse, they inhabit the same nature. The Amerindian way of thinking, by contrast, is "multinaturalist," assuming spiritual unity between humans and animals, who are all persons, although accompanied by bodily differences of physical nature, whether they have fins or feet, for example. For Westerners, there is a universal similarity between bodies, with common biological functioning, while differences between societies are based on cultural diversity. For Amerindians, culture and personhood are universally shared, whereas nature and matter differ the most (de Castro 1998: 470). Rather than cultures, it is natures that differ and are distinguished. This argument is a more fundamental challenge to Western thought than is cultural relativism. For many Amazonian peoples, and others elsewhere, "animals are people, or see themselves as persons," while the normal body of each species is a kind of clothing that "conceals an internal human form" (470–71). Societies with such beliefs are described by de Castro as *animist*. **Animism** is more commonly discussed in terms of religious ideas that locate an animate spirit in non-humans—not just animals, but also plants such as sacred trees, or non-living objects such as sacred mountains.

French anthropologist Philippe Descola, another Amazon specialist, argues that there are four different fundamental ways of understanding the relationship between humans and animals among the world's societies. Here we consider only animism and naturalism, which he says are polar opposites (Descola 2013).[3] Fundamental differences in cosmologies arise from varying treatments of two basic principles by societies. "Physicality"

---

3 The other two ontologies, which we won't discuss, are totemism, where groupings of humans and non-humans are united because they share interior as well as physical attributes, and analogism, in which humans and non-humans are made up of fragmented essences.

is the material character of our exteriors or bodies. "Interiority" concerns whether the body contains what Western thinkers variously call "the mind, the soul, or consciousness, intentionality, subjectivity, feelings, and the ability to express oneself and to dream" (Descola 2013: 116). The two terms resemble the Western distinction between the body and the mind. For naturalists, only humans have minds, while animists see animals as persons who also possess them. Under certain conditions, outer bodies can be shed to reveal a common human shape. These distinct worldviews, which are now usually called "ontologies" (ideas about the kinds of entities that exist), are seen by Descola as stable over long periods and resistant to change. He discusses how the findings of primatologists have constantly undermined claims about the unique features of humans. Yet even recognition that non-human primates use and make tools, can learn language, and transmit what they learn in manners similar to culture, are culturally managed by Western humans in ways that prevent fundamental challenges to beliefs in human exceptionalism. He also argues that increasing recognition of some (other) animals as legal persons has so far failed to erode the boundaries erected by Western naturalist thought between us and them. Despite the challenges, **naturalism** maintains its characteristic nature–culture dualism.

Such dualisms are far from universal. Different conceptions of human–animal relations were held in Europe before the Enlightenment (Descola 2013), and continue to be found elsewhere. There is far too much anthropological research on the variations in human understandings of other species to attempt a summary here. We can only offer some examples. We suggest that from a posthumanist perspective, even explaining the diversity of understandings would not go far enough. The emphasis would still remain on how humans think about our relationships with other animals, what they mean to us, how we can explain these variations, and what the consequences of particular worldviews might be. These are all important and fascinating questions, but the posthumanist perspective urges us to go even further. A post-anthropocentric anthropology needs to ask more than just how humans understand other creatures. It should also bring other species into our ethnographic explorations in ways not limited to what they mean for us. There are, of course, research challenges in doing so without surrendering to the anthropocentric assumptions of mainstream Western science, which rejects any attribution of intentions to animals as illegitimate anthropomorphism; that is, treating them as being like humans. As we discuss below, seeing pets and other animals as "fuzzy humans" is another kind of anthropocentrism that distorts our understanding of other species. Animals and plants interact with each other and their environments, with dynamics that are often independent of their relationships with humans or our understanding of their interactions. A multispecies ethnography has to incorporate interactions between forests and mushrooms, not just what that interaction means for humans (Tsing 2015).

In Chapter 2, we mentioned Michel Callon's 1986 study of the scallop industry. The complex reproductive biology of scallops constrained what kinds of fishery could be sustainable. The ANT argument that non-humans serve as actants that make a difference in networks because of their particular characteristics raises some issues for multinaturalism as discussed by de Castro and Descola. The rejection of human–animal dualism by animists is certainly useful in challenging our taken-for-granted assumptions. Yet, from a posthumanist perspective, we should consider that their treatment of animal persons is based on their own cultural understandings of what those other species need, want, or should get. Making sacrifices to scallops without an adequate understanding of how scallops "really" reproduce would be unlikely to prevent devastation of the scallop population, if animists were to control and misuse large fishing fleets and resources without adopting some of the beliefs and practices of naturalist scientists. In the Anthropocene, however much we might want to criticize them for ethnocentric dualism and their contributions to the ravaging of our environment, the natural sciences' ability to uncover knowledge of the actual needs of species and ecosystems seems indispensable to getting through the challenging coming decades and centuries without fundamentally destabilizing our shared globe.

The natural sciences are similarly flawed in their reliance on the nature–society dichotomy, however. Tim Ingold (2000: 1) says that his work has been driven by a feeling that there is something wrong with social or cultural anthropology if it ignores human beings as biological organisms, but that "there must be something equally wrong with a biological anthropology that denies" a significant "role for agency, intentionality or imagination in the direction of human affairs." He has explored ways to replace the dichotomy of nature and culture with the synergy of organism and environment, an approach that applies not only to *Homo sapiens* but also to other living beings (9). Recently he suggested the idea of "biosocial becomings" as a way to transcend the culture–biology duality, stressing that all life is both social and biological, and attempting to allow our human trajectories of becoming to be "re-woven into the fabric of organic life" (Ingold 2013: 10). With a comparable agenda, Donna Haraway (2008) refers instead to "naturecultures."

Eduardo Kohn (2013) supports the call for multinaturalism, but at the same time doesn't want to be restricted to human interpretations in going beyond multiculturalism. He begins *How Forests Think* (another Amazonian study) by relating the warning he was given that he should sleep face-up; by looking back at an intruding jaguar, he would be seen as another self rather than as prey. If the interpretations made by jaguars matter, he argues, "then anthropology cannot limit itself just to exploring how people from different societies might happen to represent them as doing so" (1). That jaguars "represent the world does not mean that they necessarily do so as we do." This changes "our understanding of the human. In that realm beyond the human, processes, such as representation, that we once thought we understood so well, that once seemed so familiar, suddenly begin to appear strange" (2). Kohn (2007) proposes an anthropology that would situate "all-too-human worlds within a larger series of processes and relationships that exceed the human" (6). Achieving this ambitious aim cannot be achieved within a "multiculturalist and dualistic framework"; instead, he urges us to look to multinaturalism (18). As with ANT, pursuing the non-human relationships that permeate human worlds requires us to open up our analyses, pursuing a study of "open wholes." Kohn (2013: 39) is very relevant to holism without boundaries. The study of open wholes requires including non-human living beings and recognizing that they have their own dynamics ("material agency," in Pickering's term) that transcend our beliefs about them and interpretations of them.

Kohn (2013: 182) says that humans do not simply impose form on the tropical forest, but rather the forest forms its nature "in myriad directions thanks to the ways in which its many kinds of selves interrelate." Forests, and other ecologies, have coevolved as communities, and how they respond to our interventions depends fundamentally on the history of these interdependencies. For example, Anna Tsing (2015) examines the interactions between forests, matsutake mushrooms, and the humans that profit from picking and selling these extremely expensive delicacies. These mushrooms do not thrive in rich soil and are usually best found in secondary forests, abandoned after heavy forestry or other conditions (such as glaciers) have destroyed the primary forest and left rocky and unproductive soil. The synergy between pines and mushrooms benefits both, and also provides a new industry for areas where forest companies have moved on and left communities and lives in ruins.

Why the Non-Humans Disappeared from Anthropology

It is important to understand why non-humans have had to be brought back into anthropology by scholars like Ingold, Descola, and Kohn. To answer this, we must understand why they faded from view in the first place. Anthropology in the first half of the twentieth century or so more commonly included multiple species in their ethnography than did the work of anthropologists who were strongly influenced by interpretive and postmodern anthropology from the 1970s on. These influences continue, even if few anthropologists now describe themselves as postmodernists. To explain these distinctions between different anthropological methods and perspectives, we have to make a detour through some debates in the discipline, but we will keep it short and focus on issues important for posthumanism.

It is not surprising that classic ethnographies were more likely to include non-humans in their writings. At the time, there was little reliable and unbiased information about the small-scale societies that lacked writing before colonization. The first task was to describe their ways of life as holistically and completely as possible. This task was particularly important for studies of indigenous populations in North America and Australia, because their cultures and natures were rapidly being destroyed and transformed as part of what has been described as "cultural genocide." American cultural anthropology during this period was often described as "salvage anthropology": collecting knowledge from elders before it was lost. Almost no one else was doing research with these groups at the time, so there was no pressure to specialize, unlike sociologists who had to distinguish their field of study from those of economists, political scientists, agronomists, and so on. Studying crops, livestock, hunting, and local understandings of non-humans was not driven by theoretical considerations like those we discussed in the first half of this chapter, but by the requirements of producing knowledge, researching, and writing a total ethnography of a people's way of life. Since these non-human entities were important to the people, they properly came to be the subject of ethnographic study.

With the rapid expansion of universities after World War II, many more professors were employed. Although anthropologists were a very small part of the total, this resulted in a dramatic expansion of what had been a very small circle of researchers in the past. With this growth of the discipline, graduate students started to do their research among groups that had already been studied. In doing so, they tended to specialize in particular topics such as religion, political organization, kinship, and so on. Holism was still practiced, but primarily to know enough about the group in general to be able to understand how, for example, the political organization was influenced by kinship and religion and how the people made a living. This specialization made theoretical innovation and fashion more important if an anthropologist wanted to get attention beyond his or her field of specialization. This situation facilitated the adoption of postmodernism.

The characteristics and consequences of **postmodernism** in anthropology have been quite different from those of postmodernism in other fields and disciplines (Smart 2011). Defined in widely different ways, we find the most useful feature of postmodernism to be its rejection of modernity, which is basically the Western Enlightenment principles of reason, progress, and the improvability, if not perfectibility, of humans. Postmodernism saw these as "grand narratives" that were no longer supportable. Those principles were both Eurocentric and no longer convincing in a world where most people lived in countries that had been subjected to colonialism in the name of progress.

Postmodernism can be separated into post-structuralist and humanist strains. Most anthropologists are in the latter camp. Postmodernism in the larger sense is often seen as including authors such as Derrida and Foucault. These authors, in fact, are better understood as post-structuralists and as critics of humanism. In anthropology, however, the genealogy from interpretive and symbolic anthropology to postmodernism was mostly humanist. **Post-structuralism** in the forms developed by Derrida, Foucault, Haraway, Althusser, Latour, and Deleuze and Guattari, among others, questions the coherence of the individual. People are instead seen as intersections, or assemblages, of relations, contexts, or actor-networks; in other words, outside forces congregate in individuals that are formed as particular kinds of people. Post-structuralism rejects humanist assumptions about rational and autonomous Man and its grand narratives of Enlightenment and progress. This critique of the person raises problems particularly for anthropology, but the same difficulties are presented for postcolonialism.

Postmodern anthropology owed great debts to the interpretive approach championed by Clifford Geertz, an anthropologist who conducted research in Indonesia and Morocco. In his pivotal volume of essays, *The Interpretation of Cultures* (Geertz 1973: 10), Geertz presented culture as an "ensemble of texts," which formed a tangled web within which people construct meaning. The anthropologist attempts to write ethnographies through interpreting the interpretations of the people he or she does research with. The analyst thus needs to pay close attention to the various voices of people and how they are presented in their publications.

In part because of Geertz's influence, as well as our discipline's legacy of working in oppressive colonial contexts, postmodern anthropologists tend to be very humanist. There were only two passing mentions of post-structuralism in the foundational books for postmodernism, *Writing Culture* (Clifford and Marcus 1986) and *Anthropology as Cultural Critique* (Marcus and Fischer 1986). To an extent perhaps greater than in other disciplines, texts became increasingly central to anthropological discussions. Even social interactions and behaviors were thought to be most usefully interpreted as if they were texts. From a posthumanist perspective, the problem is that animals are never the authors of texts but only the objects of human texts, talk, and interpretation of their meaning for humans. Olga Solomon (2012) notes that there has recently been a "counter-linguistic turn" in animal studies and disability studies that tries to identify forms of subjectivity that are not language based.

One of the things that postmodern anthropologists do extremely well is present narratives that open windows to radically different worldviews. They encourage disparate voices, and in documenting them, they contribute to our future knowledge of our present, including cultural variation within particular cultures. Postmodern anthropology promoted listening to and including alternative, often marginalized voices, certainly a good thing to do, and something we would recommend to all politicians and government officers, in particular. Paying attention to **polyvocality** (also referred to as multivocality) is an attempt to recuperate more of the voices encountered in our research, giving expression to them in our writings rather than privileging the researcher's perspective. While worthy and productive, this research focus appears to exclude voiceless non-humans. How should we give voice to non-humans, or are we limited to trying to speak *for* them? Anthropocentrism is still operating in postmodern anthropology, whereas the post-structuralist position tends to adopt the position that we perform our parts, or voice our voices, through chains of social and material relations about which we have only dim awareness. Anthropological postmodernism tends to resemble postmodern architecture, which Scott Lash (1990) argued was much more humanist than its modernist predecessors, in promoting (if rarely achieving) the anthropomorphism, anthropocentrism, and anthropometrism largely rejected by modernist architects, who stressed function over form and rejected ornamentation.

Strengths that anthropology once had, we suggest, are undermined by the postmodernist/materialist, science/antiscience divides that split anthropology in the last quarter of the twentieth century and were part of what were called the "Science Wars" that widely affected academia. Divisions between the four fields might be less threatening, and our complementarities better valued and fostered, if we could encourage a less anthropocentric and humanist perspective within social and cultural anthropology. We would also need more engagement in social theory by biological anthropologists. If English professors can find esoteric debates in biology to be stimulating fodder for re-visioning their practices, surely anthropologists can benefit from greater conversation with our colleagues down the hall. This book is in part a tribute to our good luck in being part of a department of anthropology where such mutual benefit was recognized and preserved, with the two main streams in our department both emphasizing social relations among people (social and cultural anthropology) and non-human primates (formerly primatology, now relabeled as biological anthropology) and respecting what the other stream contributed to knowledge. Such mutual respect was not common in anthropology departments after the advent of postmodernism, which made it more difficult to build on the holistic tradition of anthropology as a discipline with four perspectives on what it meant to be human: the social/cultural, biological, archaeological, and linguistic fields of anthropology, with applied anthropology sometimes added as a fifth field.

Postcolonialism shares with anthropology fundamental awkward questions about posthumanism. Are we not just pulling the rug of individual coherence and species importance out from under the feet of peoples who, only in the last century or so, have regained self-determination? Descendants of populations subject to expropriation, displacement, and genocide—that is, all those whose predecessors lived in colonized places—can be forgiven if they reject posthumanist ideas for trying to strip away something that was achieved only in the decades after World War II, and often still only partially. If people are seen as only contingent assemblages of wider forces, as post-structuralists suggest, what does that imply for the freedom and self-determination that they have achieved against resistance from their oppressors and exploiters? Postcolonialism and postmodern anthropology find the theoretical arguments for the rejection of anthropocentrism a challenge in relation to their demand for equal space for non-Western subjects as resisting and transformative agents and knowers. After struggling for civil and human rights, it seems tragic to find the subjectivity they have achieved being dissolved in networks or webs of external forces.

Postcolonial scholars who work with indigenous populations, such as the First Nations of Canada, have found posthumanist ideas particularly difficult. Danielle DiNovelli-Lang (2013) notes that posthumanist scholars often find themselves having to deconstruct, or analytically challenge, ideas of indigeneity, and instead adopting Eurocentric theoretical perspectives, even as they critique Enlightenment ideas. She asks whether it is possible to combine posthumanist and postcolonial theories, or if the posthumanist erasing of the boundary between nature and culture instead erects new divisions between peoples, requiring the maintenance of the humanism in postcolonial approaches. This kind of risk can be seen in the dualistic distinctions between naturalist and animist ontologies presented by Descola, which have the potential to reinforce stereotypes about other cultures. This risk is intensified by his belief in the long-term stability and internal unity of such ontologies (i.e., differences between natural scientists and environmentalism are seen by him as simply variant forms of naturalism). It is beyond the scope of this book to try to resolve these theoretical and political issues. Our point here is simply to stress the way in which both anthropology and postcolonial studies have political and conceptual difficulties in adopting a post-anthropocentric path—difficulties that do not apply with the same force to scholars in most other fields.

Returning to postmodernism, we have argued that meanings and voices were central in the approach, and its key issues revolved around writing and representation. Nature, production, and material culture were sidelined from the analysis, to a much greater extent than in earlier realist ethnographies (Knauft 1996: 23). Even fighting roosters are analyzed by Clifford Geertz as metaphors used by humans, as symbols that tell us what people think about the animals and about other people rather than actants with their own material agency.

Geertz's essay about cockfights in Bali is one of the most influential anthropological publications ever written. It has been widely praised outside anthropology, in part because it is written with vibrancy and elegance, and without jargon, yet provides a clear account of what anthropology is about. The essay dissects the local preoccupation with illegal cockfights in a way that reveals multiple layers of meaning and the underlying tension between the normally aloof and self-composed Balinese and the underlying emotions that, in ordinary circumstances, need to be contained. The cockfight "brings together themes—animal savagery, male narcissism, opponent gambling, status rivalry, mass excitement, blood sacrifice—whose main connection is their involvement with rage and the fear of rage, and, binding them into a set of rules which at once contains them and allows them play" (Geertz 1973: 27). Roosters, and their struggles with each other, are metaphors for the underlying animality of people who pride themselves on their self-mastery. According to Geertz (1973), the Balinese see animality as the "direct inversion, aesthetically, morally, and metaphysically, of human status" (6). They reject any animal-like behavior to the extent that babies are not allowed to crawl. Even eating is seen as "a disgusting, almost obscene activity ... because of its association with animality" (7).

In part because of its prominence, the essay has received a great deal of criticism. One of the more influential critiques was that of William Roseberry (1982). Roseberry points out that the essay manages to omit three important aspects of Bali from the interpretation: gender, the political processes of colonialism and nationalism, and inherited caste status. Yet Roseberry himself fails to point out another gap that is important for the purposes of this book: we learn nothing about the role of chickens in Balinese economy and ecology. Bali has a densely occupied rural landscape, with steep hillsides that are intensively terraced for rice. Chickens are presumably important for other things than fights, yet their role in Balinese nutrition is not mentioned, nor is the possibility that their waste is useful as fertilizer, among other potential utilities. As well as being discussed almost exclusively as metaphors for textual construction, chickens are dematerialized and separated from the hens, eggs, and sources of food that clearly must also be involved. This gap is particularly surprising in relation to Geertz's (1963) earlier work, which includes a well-researched book on the ecological anthropology of Java, another Indonesian island.

Human–Animal Studies

Our lives, societies, and cultures are inseparable from non-human life; yet, until recently, the core of cultural and social anthropology has had less to say about these issues than our predecessors before the postmodern turn. Close consideration of such issues had been left to less-fashionable and more-specialist fields that include a stronger dose of materialism, such as ecological anthropology, political economy, and medical anthropology. However, the last decade has seen a rapid growth of what we are calling multispecies ethnography. We have chosen to adopt the label "multispecies ethnography" because of the useful way in which it destabilizes the anthropocentrism that persists in the label of "human–animal relations" (Kirksey and Helmreich 2010), also known as **anthrozoology**. Multispecies ethnography acknowledges that interactions between different non-human species are not necessarily mediated (only) through their interaction with humans. Donna Haraway (2008), for instance, traces the three-way interaction between sheep, guardian dogs, and humans. Anthrozoology also leaves microorganisms and plants out of the discussion. For example, the interaction between humans and cattle in East Africa was deeply influenced by the sleeping sickness spread by tsetse flies (note that research is being conducted on using gene editing, discussed in Chapter 4, to give African cattle immunity to sleeping sickness). The other crucial advantage of thinking about multispecies ethnography is the emphasis it places on ethnographic research.

There is a vast and rapidly expanding body of work on human–animal studies (see De Mello 2012 for a readable overview) and critical animal studies (Calarco 2015), and space does not permit any attempt to summarize here the breadth of such work and its implications. We prefer to emphasize multispecies ethnography because of its more inclusive and less Eurocentric emphasis. In addition, while human–animal studies research in other disciplines is often theoretically exciting, there is a tendency for speculation and theoretical jargon to predominate at the expense of solid ethnographic research. Fieldwork—not just interviews, but spending substantial amounts of time in the places where people live their lives and interact with their non-human companions—helps illuminate the interaction between different species and opens us up to the potential of surprise in what we find. Emphasizing ethnographic research that incorporates more than just our own species links us back to the pre-postmodern moment in anthropology when animals and plants were central elements of ethnographic research—when cattle, pigs, yams, rice, and so on were thoroughly embedded and attended to in almost every major anthropological monograph.

Given the heavy emphasis in the first part of this chapter on work in Amazonia, we shift now to research and issues that are closer to the experience of most university students. Michael Moore's satiric documentary *Roger and Me* (1989) on the human consequences of the decline of the car industry in Flint, Michigan, provides an apt slogan for this discussion. In an infamous segment, Moore interviews a laid-off auto worker who has put up a hand-lettered sign for her money-making venture: "Rabbits or bunnies, pets or meat." Taking inspiration from these choices of label and material outcome, in the remainder of this chapter we concentrate on our interactions with animals as both pets and food.

Domestication

The **domestication** of animals and plants changed those species dramatically, with many agricultural crops now unable to reproduce without human intervention. But it also transformed humans. The archaeologist V. Gordon Childe (1936), who coined the term "Neolithic Revolution," titled his book *Man Makes Himself.* The Neolithic Revolution was a pivotal transformation in history, in which formerly mobile humans began a farming way of life, accompanied by a sedentary lifestyle and population increases. His book title is clearly anthropocentric (as well as exclusionary to females) and presents animals and plants as passive in the process (O'Connor 1997). Other scholars reject the idea of domestication as one-sided. People did not by themselves bring dogs, sheep, or other animals into domestication; rather, they entered into a new interaction by behavioral adaptation on the part of both species—what Pickering would call a dance of resistance and accommodation. Since many species were more resistant to domestication, we might think of a ballroom filled with more or less eager partners. The domestication dance contributes to behavioral coevolution, through which mutualistic relationships developed because the interaction was to the benefit of both species (Haraway 2003: 32). For example, one of the consequences of the domestication of wild canids (the taxonomic family that includes dogs, wolves, foxes, and so on) into dogs has been a reduced size (until giant breeds were produced by artificial selection). Perhaps instead of people selecting smaller animals for rearing, it may have been the smaller ones that actively sought a closer relationship with humans for protection (O'Connor 1997). Peter Bleed (2006: 8) takes this idea one step further when he suggests that the Neolithic expansion of human populations created an ecological niche that could be "occupied" by other species. He argues that we should not look at domesticates as "passive resources or even as co-evolving species, but as influential occupants of a dynamic set of opportunities afforded by people" (9). A recognition that human populations can be exploitable niches applies, of course, even more aptly to zoonotic diseases. Human destruction of habitats increases the attractiveness of our own species as a new frontier for viruses whose usual hosts are a dwindling niche.

Turning to crops, Anna Tsing (2012: 145) tells us that "cereals domesticated humans." The cultivation of grains did not make life easier for foragers, but it did do so for the states and elites that emerged with it. They encouraged sedentary, stable farms that could be controlled and from which surpluses could be extracted. This new lifestyle changed everything from fertility to warfare (Tsing 2012). Particular kinds of staple grain production encouraged different kinds of social and political organization.

Dogs have been associated with humans for longer than any other animal—perhaps for 100,000 years, but at least for 17,000 (Udell et al. 2010). Many scholars suggest that domestication produced not only substantial bodily changes but also genetic changes that heightened canine sensitivity to human social cues, their ability to "read" our behavior and intentions. Domestication is usually seen as a genetic change due to both natural and artificial selection. It has been distinguished from taming, a process that occurs during an individual's development; wild animals can be tamed, while domesticated animals can become feral and untamed. Monique Udell and her colleagues (2010) tested the domestication hypothesis that genetic change is the main cause of a canid's sensitivity to human behavior. They found that if wolves are fully socialized by humans from a young age, they do as well or better than dogs in learning to respond to a human pointing with their arm. They conclude that rearing, or nurture, is more important to dogs' sensitivity to humans than is genetic change. This is not to say that millennia of domestication do not make a difference. Experiments with foxes, in which breeding was used to select for tameness, were able to produce changes in only a few generations. Some of the physical changes associated with domestication (such as shorter tails and floppy ears incapable of the usual social signaling) occurred without being selected for.

Petting Animals

In *The Companion Species Manifesto*, Donna Haraway (2003) challenges us to understand how much of ourselves is a result of relationships with other species and to take seriously our encounters with "significant others," like our dogs. From a critical feminist position, she sees humans as the products of how they relate to other people, creatures, and things. Although she explores ethnographically the deep emotional ties that develop between people and dogs, she is critical of the strong contemporary tendency to see dogs as "furry children," which misinterprets and neglects the great differences between the species. It is the encounter with something different that offers the most potential for learning and expansion of our ways of thinking about the world, rather than squeezing animals into pre-existing human social and kinship roles (Haraway 2003: 37; see also Lorimer 2015). While she stresses that we do incorporate dogs into our families, she rejects "all the names of human kin for these dogs," arguing that we need new ways to specify these deepening relationships "in non-humanist terms in which specific difference is at least as crucial as continuities and similarities across kinds" (Haraway 2008: 67).

Recognition of fundamental commonalities simultaneous with differences opens greater potentials for flourishing together on the basis of distinctive but coevolved and compatible **naturecultures** that have linked our histories together. She uses the example of Australian shepherds, whose creation as a canine breed is tied up with global history, agricultural landscapes, and technological changes. By learning about the history of her own dog, she found herself drawn into learning about entanglements between animals, humans, and technologies. These linkages have become even more complex. Dogs have become an ever larger market for consumer goods and veterinary services, and dogs have new productive roles as therapists and guides (Coulter 2016), to add to reinvigorated ones, such as guardians for sheep, seen as an environmentally friendly way to control grass on ski hills and golf courses.

Despite Haraway's concerns, recent decades have seen a rapid growth of treatment of dogs and cats as "furry children." Eighty-three per cent of Americans refer to themselves as the "mum" or "dad" of their cats or dogs, up from 55 per cent 20 years earlier. Over 90 per cent consider their dog or cat to be a family member. Spending on companion animals increased by 2.5 times from 2000 to 2013, to an annual $55 billion. Courts have started to award damages for mental suffering and loss of companionship to the owners of slain pets, previously reserved for the wrongful death of kin (Grimm 2014).

This trend is not restricted to North America. In Japan, skyrocketing pet ownership has combined with rapid aging and dropping fertility rates to create a situation where pets outnumber children under the age of 15. Paul Hansen (2013) did research in Osaka by walking a friend's dog to dog parks, where they became the "best field assistants," sparking conversations among otherwise aloof people. While people may follow the conventions of civil inattention to others in urban spaces, dogs do not, and this drags people together. In informal conversations, followed by a questionnaire he asked people to mail back to him, he found that they unanimously considered their dogs to be family members. For many people, at least when it comes to companion animals, anthropocentrism has been replaced, or supplemented, by **anthropomorphism**, the ascription of human qualities to non-humans. Usually considered a flawed kind of reasoning, scholars in primatology (de Waal 2008) and other fields increasingly argue that a cautious anthropomorphism is a necessary correction to anthropocentrism because it helps us to consider non-humans as having agency and intentions (Bennett 2009; Hansen 2013).

Eating Animals

Cary Wolfe (2012: 54) emphasizes that, at the very time when some animals are being seen as members of our families and communities, billions of animals in factory farms—many "of whom are very near to or indeed exceed cats and dogs ... in the capacities we take to be relevant" to social standing, such as emotional bonds and complex social interactions—have "as horrible a life as one could imagine" in confined feeding operations in industrial agriculture. How do we reconcile these polarized treatments of animals that are central to our lives in very different ways, as loved ones and ones we love to eat?

Melanie Joy (2009: 14) argues that this distinction is related to the **cognitive schemas** that we possess and which deeply influence how we behave. An animal can be classified as "prey, predator, pest, pet, or food. How we classify an animal, in turn, determines how we relate to it—whether we hunt it, flee from it, exterminate it, love it, or eat it.... When it comes to meat, most animals are either food, or not food." When you eat a hamburger, most people don't envision a cow, but if you were told that your dinner was a golden retriever, the resulting images and thoughts would probably produce a sense of disgust and prevent most North Americans from being able or willing to eat it (Joy 2009: 11). Her conclusion, that eating any kind of animal requires some degree of psychic numbing, is useful for our understanding of Western consumers, but her analysis could benefit from exposure to the work of anthropologists on hunting in other cultural contexts. Ingold (2000: 13) says that the Cree people of northeastern Canada believe that caribou offer themselves at the moment of the kill "intentionally and in a spirit of good-will or even love toward the hunter. The bodily substance of the caribou is not taken, it is received. And it is at the moment of encounter, when the animal stands its ground and looks the hunter in the eye, that the offering is made." Rather than "psychic numbing," this represents a completely different worldview with much richer relations between humans and the non-human persons that they eat.

Marshall Sahlins (1978: 176) offers a fascinating explanation of American revulsion against the idea of eating dogs, cats, and horses. He suggests that these food taboos can be seen as an elaborated system of avoidance of symbolic cannibalism. Companion animals participate in American society as subjects who have personal names, with whom we converse and have deep social bonds. The closer an animal is perceived to be to people, the more that eating it seems like a kind of cannibalism (at least within this cultural framework). Sahlins's model goes further, however. The easier it is to see a resemblance between a piece of meat and a human body part, the more likely it is for people to feel uncomfortable eating it. This revulsion is reflected in prices: kidneys and livers have lower prices than sirloin or T-bone steaks, which most consumers could not map onto the cattle carcass. Breaded chicken nuggets and hamburger meat seem even more distant from a living animal.

In any cultural context, food elicits reactions of desire and disgust; it may also bring with it either revitalization or poisoning. Food can prompt memories of home and loved ones, or reactions of horror at the culinary choices of others, such as in efforts to ban the consumption of dogs, cattle, or horses (Sahlins 1978). None of these reactions are purely biological, nor can they be fully separated from our biological natures. Taste includes both biochemical reactions and attitudes to those reactions. Both are culturally constructed and modified, but their physicality is also relevant to food preferences, since malnutrition often creates cravings for particular kinds of food. Whether food entices us or disgusts us depends on our cultural context and bodily habits. Despite this visceral conservatism of our tastes, diets can change in complex and poorly understood ways. Sushi is a good example (Bestor 2000). The idea of eating raw fish still unsettles many North Americans, yet its consumption has also grown dramatically. It is particularly difficult to deny the importance of embodied response when we consider food. The inherently visceral nature of eating is tied up with culturally derived feelings of pleasure and disgust (Miller 1997; Macbeth and Lawry 1997).

For many people, animal protein represents desires for a more prosperous life, and the rapid increase of its consumption in emerging economies is having a dramatic impact on the world's agricultural system and environment. The declining global reserves of grain and its associated rising prices in the last decade are attributable in substantial part to the feeding of animals for increased meat consumption in countries like China, India, and Brazil. According to some researchers, the only sustainable and significantly expandable sources for non–plant-based proteins are jellyfish and insects. The "yuck factor" is a very powerful deterrent, at least in the West, to widespread adoption of insectivorism, despite its ecological merits.

Eating insects seems to be on the verge of becoming more mainstream, though, as the recent appearance of items such as scorpion pizza as novelty festival food suggests. It is increasingly seen by advocates as a green alternative to the inefficient and environmentally damaging beef, pork, and chicken industries. But there is obviously a huge amount of Western consumer resistance to be overcome. A team of four postgraduate students from the Royal College of Art and Imperial College London, who wanted to tackle the growing issue of food supply in an increasingly hungry world, developed a prototype brand, "Ento," as a "roadmap for introducing edible insects to the Western diet.... Motivated by the failings of the livestock industry, as well as the environmental and nutritional benefits of insects, the team wanted to see how this provocative new food source could be introduced to Western diets. The project is about driving cultural change through understanding human perceptions, using strategic design thinking, as well as through creating innovative and compelling experiences" (Core Jr. 2012). The Ento team drew inspiration from the Western adoption of sushi, and the abstract forms into which it is processed (small cubes, for example, as they found that the more the food could be recognized as insectoid, the more resistance was evidenced). Elegant packaging was intended to appeal to trendsetting foodies who could then influence more mainstream consumers, as has happened with sushi.

A key feature of posthumanism, we argue, is the rejection of anthropocentrism, and the recognition that the very nature of being human involves our entanglement with, and reliance on, non-humans. While we evolved through both eating and being eaten by other species, in recent history the relations between prey and predator have become predominantly one-way, at least if we neglect the parasitism of our bodies. How we eat is no longer only an issue of survival or of cultural and aesthetic preferences; in the Anthropocene, we have new responsibilities to a myriad of species that will be influenced by our choices—not just what and how we eat, but more generally how we live, and with which other companion species (Lorimer 2015). With a return to the greater species inclusivity of classic ethnographic research, enhanced by the theoretical and methodological advances since, anthropologists could contribute even more to our knowledge of the web of more-than-human relationships. In the next chapter, we go beyond life forms to consider our relationships with tools and technology.

Discussion and Activities

Consider how you classify or categorize a non-human animal that you have interacted with. Does its classification create any conceptual confusion for you? How does this classification affect your behavior toward this animal (or vice versa)? Do any of the issues discussed in this chapter challenge your classification and your understanding of the relationship?

Consider the kinds of things that you would not be willing to eat. How can you understand these food restrictions? Are they part of a cultural tradition that you have adopted without consideration, or are some of them the result of explicit religious, ethical, or health decisions?

Do a Google search for edible insects. Which of these would you be most willing to eat, which most reluctant? Why?

Google the term "fawn napping." Discuss its implications for our relationships with "wild animals," and how technology can influence changes in these relationships. What other technologies are affecting our relationships with other animals?

Compare attitudes toward marine mammals in the popular films *Flipper* (1964) and *Free Willy* (1993), and the documentaries *Blackfish* (2013) and *The Cove* (2009).

The German Shepherd Rin Tin Tin received the most votes for best actor at the first Academy Awards in 1929. The Academy wished to appear more serious, so Rin Tin Tin was removed as a choice and the votes were cast once more. Discuss what this means for animal–human relations, and whether you think that the Oscars should be less speciesist.

Additional Readings and Films

Readings

Armstrong, Susan, and Richard G. Botzler, eds. 2008. *The Animal Ethics Reader*. New York: Routledge.

Coulter, Kendra. 2016. *Animals, Work, and the Promise of Interspecies Solidarity*. New York: Palgrave MacMillan.

Hurn, Samantha. 2012. *Humans and Other Animals: Cross-cultural Perspectives on Human–Animal Interactions*. London: Pluto Press.

Pollan, Michael. 2006. *The Omnivore's Dilemma: A Natural History of Four Meals*. New York: Penguin.

Singer, Peter, ed. 2013. *In Defense of Animals: The Second Wave*. New York: John Wiley & Sons.

Wolfe, Cary. 2012. *Before the Law: Humans and Other Animals in a Biopolitical Frame*. Chicago: University of Chicago Press.

Films

Bambi. 1942. Film directed by David Hand. Burbank, CA: Walt Disney Productions.

Black Beauty. 1946. Film directed by Max Nosseck. California: Edward L. Alperson Productions.

Black Beauty. 1994. Film directed by Caroline Thompson. Burbank, CA: Warner Bros.

Dogs with Jobs. 2000–2004. TV series directed by Serge Marcil et al. Montreal: Cineflix.

Hachi: A Dog's Tale. 1987. Film directed by Seijirô Kôyama. Japan: Mitsui Company Ltd.

Hachi: A Dog's Tale. 2009. Film directed by Lasse Hallström. Culver City, CA: Stage 6 Films.

Homeward Bound: The Incredible Journey. 1993. Film directed by Duwayne Dunham. Burbank, CA: Walt Disney Pictures.

The Horse Whisperer. 1998. Film directed by Robert Redford. Burbank, CA: Touchstone Pictures.

The Jungle Book. 1967. Film directed by Wolfgang Reitherman. Burbank, CA: Walt Disney Productions.

The Jungle Book. 2016. Film directed by Jon Favreau. Burbank, CA: Walt Disney Pictures.

Lassie Come Home. 1943. Film directed by Fred M. Wilcox. Beverly Hills: Metro-Goldwyn-Mayer.

Marley & Me. 2008. Film directed by David Frankel. Los Angeles: Fox 2000 Pictures.

Max. 2015. Film directed by Boaz Yakin. Beverly Hills: Metro-Goldwyn-Mayer.

My Dog Skip. 2003. Film directed by Jay Russell. Los Angeles: Alcon Entertainment.

Old Yeller. 1957. Film directed by Robert Stevenson. Burbank, CA: Walt Disney Productions.

Seabiscuit. 2003. Film directed by Gary Ross. Los Angeles: Spyglass Entertainment.

Secretariat. 2010. Film directed by Randall Wallace. Burbank, CA: Walt Disney Pictures.

The Selfish Giant. 2013. Film directed by Clio Barnard. London: Moonspun Films.

The Story of the Weeping Camel. 2003. Film directed by Byambasuren Davaa. Mongolia: Mongolkina.

Tulpan. 2008. Film directed by Sergey Dvortsevoy. Germany: Pallas Films.

War Horse. 2011. Film directed by Steven Spielberg. California: DreamWorks Pictures.

Documentaries

The Botany of Desire. 2009. Documentary directed by Michael Schwarz and Edward Gray. Menlo Park, CA: Kikim Media.

Fast Food Nation. 2006. Documentary directed by Richard Linklater. Century City, CA: Fox Searchlight Pictures.

Fed Up. 2014. Documentary directed by Stephanie Soechtig. Santa Monica, CA: Atlas Films.

Food Fight. 2008. Documentary directed by Christopher Taylor. United States: November Films.

Food Inc. 2008. Documentary directed by Robert Kenner. New York: Magnolia Pictures.

Ethnographic Films

The Basques of Santazi. 1987. Film directed by Leslie Woodhead. London: Granada Television.

Corn and the Origins of Settled Life in Meso-America. 1964. Film directed by Jack Churchill. Boston: Educational Development Center.

Dani Sweet Potatoes. 1974. Film directed by Karl G. Heider. Boston: Educational Development Center Inc.

Dead Birds. 1963. Film directed by Robert Gardner. Boston: Peabody Museum.

Himalayan Herders. 1977. Film directed by John and Naomi Bishop. Watertown, MA: Documentary Educational Resources.

The Holy Ghost People. 1967. Film directed by Peter Adair. United States: Thistle Films.

In Search of Cool Ground: The Kwegu. 1985. Film directed by Leslie Woodward and David Turton. London: Royal Anthropological Institute.

The Nuer. 1971. Film directed by Hilary Harris and Robert Gardner. Cambridge, MA: Film Study Center.

Ongka's Big Moka. 1976. Film directed by Charlie Nairn. London: Granada Television International.

If you are looking for films on particular kinds of animals, see *Zooscope: Animals in Film Archive* at https://zooscope.english.shef.ac.uk/articles/advanced_search.

CHAPTER 4

TECHNOLOGY, CYBORGS, AND TRANSHUMANISM

In this chapter, we focus on technology. In doing so, we emphasize one dimension of holism in anthropology—namely, its breadth of coverage. Anthropologists tend to focus in one of four fields: social and cultural anthropology, biological anthropology, archaeology, or linguistics. Bringing these different approaches together allows a wider variety of insights into the entanglement of people and their tools.

What does it mean for this classic vision of holism within our discipline if we reinvent holism without boundaries? If there are no boundaries to limit the pursuit of knowledge and answers, does our inclusion of four fields in just one discipline make any difference, or is it just a legacy of modernist ideas of the university? Does it perhaps exist only for the narrow interests of academic departments that want to get a bigger share of the funding pie? We suggest that a post-anthropocentric anthropology should not confine research to narrow disciplinary definitions of what anthropologists can and should do. To understand a controversy, issue, or outcome, we need to learn whatever helps us to make sense of the situation.

Science studies conducted only by anthropologists would be less rich and fruitful. Still, disciplines are useful; they provide the concepts and tools needed to conduct novel research and scholarship. Disciplines provide a focus, and they tie us to legacies of ideas and findings that continue to be useful today. We stressed one of those legacies in the last chapter—that is, the inclusion of non-humans in classic ethnographies. Looking back, then, can help us to think about the way forward. If we don't know the path that we have been following, we will find it hard to get to where we are bound.

Tools and Being (Post)Human

We organize our discussion of the interaction between humans and our tools historically, beginning with the era when we became human through our collaboration with non-humans—particularly microbes, previously discussed, and fire, which we turn to below. After introducing some ideas with which to think about our interface with technology, we provide several examples of the insights arising from ethnographic research on emerging technologies, such as new reproductive technologies and smart homes, which make our living environments responsive and globally connected. The chapter ends with a consideration of the transhumanist movement, and some transhumanist visions for futures beyond the merely human. During the lifetimes of students reading this book, it seems likely that humans will be greatly transformed, much more so than in the last three decades, through their use of implanted and wearable technology. Connections with global webs of intelligences, both machine and human, will make possible very new forms of being, as they enable us to become virtually what we desire to be: human, robot, or animal (Boellstorff 2008). Yet techno-utopias have been promised before. Anthropological research on the effects of introducing technologies that could potentially radically change workplaces and economies has often found that they are implemented in ways that are compatible with prevailing ways of doing things. Their disruptive effects may come much later, if at all. Innovation becomes domesticated in various ways in different societies and contexts, as it is adopted and becomes routine practice.

We concentrate on the processes by which we interact with tools and technology, and how such interactions have changed us in fundamental ways. Humans and technology have mutually formed each other, in a millennia-long dance of agency. Tools have changed us in our adaptation to their use (Franklin 2013). Technology helps us to inhabit our worlds, and we depend on it for our survival. Yet the technology that we have created is now so powerful that many fear it might overwhelm us with its capabilities, perhaps like Big Brother in *1984* or Skynet in the *Terminator* movies. Popular articles and blogs about smart homes worry that artificial intelligence (AI) capabilities might be turned against us, as occurred with the spaceship controlled by the HAL computer in the science fiction movie *2001: A Space Odyssey*. Smart homes are the private environments that, as part of smart city projects, are becoming infused with connections, controls, surveillance, and new capabilities for their tenants. We are offered the freedom of feeding our pets treats while we are away and watching their reactions in real time. We can Skype with the other members of our household as well and even do it on a screen attached to the refrigerator. We can remotely check the expiry date on the milk in our fridge.

How should we think about homes that we will be in dialogue with? Will we consider talking with our homes to be a kind of genuine sociality, as many of us experience with our pets?

In our consideration of smart homes, we examine the transformation of one of the basic technologies that made it possible for *Homo sapiens* to survive long enough to develop the tools that have led to the Anthropocene. In places like Calgary, Alberta, Canada, where we live and teach, without shelter it would be almost impossible to last a winter, with lows below −40°C. Among other important functions, homes are tools for dealing with the elements, which, in much of the world, are not kind to the human body. The kinds of technologies we inhabit makes a difference to our lives and the way we see the world. From cave to digital hub, it's been a long and strange road we have followed, with no clear end in sight. Will we be able to become transhuman without merging with our homes, as they become even more indispensable life support systems?

Technological Futurism and Transhumanism

Living in smart homes might be much easier if we could control them with our thoughts rather than a complicated and glitchy remote control. As we were writing this in September 2015, a San Diego company began shipping the cheapest and smallest **brain–computer interface** to date, a $129 US unit worn like a Bluetooth headset. The previous month, Korean and German scientists had announced a brain–computer control interface for a lower-limb exoskeleton that decodes specific signals from within the user's brain. Attempts to develop versions to reduce mobility problems for seniors were heralded by *Zoomer*, a Canadian magazine for seniors, as opening the possibility for a future of "robo-grannies and gramps bent on world domination" (Crisolago 2015). Are these the early signs of what transhumanists such as Ray Kurzweil (2005) have described as the **Singularity**?

For Kurzweil (2005: 9), the Singularity is being produced by exponential speeding up of technological change. Within the next few decades, the result will be "the merger of our biological thinking and existence with our technology, resulting in a world that is still human but that transcends our biological roots. There will be no distinction, post-Singularity, between humans and machine or between physical and virtual reality." These ideas were popularized in the 2014 film *Transcendence*, in which a dying researcher uploaded his mind into a computer program. It was given a typical Hollywood Frankenstein's-monster plot turn rather than expressing the boundless optimism of Kurzweil and many other transhumanists. This would fit one of Joel Garreau's (2006) two dominant scenarios for a transhuman future: Heaven or Hell. "Hell" could be the result of one or more potential nightmares that lead to a catastrophic end. The other scenario is "Heaven," proclaimed by proponents who see accelerating technology as offering solutions to all our challenges through becoming something beyond human.

Garreau (2006: 209) prefers a third scenario, "Prevail," in which humans prevail, due to pure "human cussedness" and our tendency to muddle through rather than go fully in one direction or another. This position is perhaps reflected in the 2015 film *Chappie*, in which experiments on a police robot in a future South Africa lead to superhuman intelligence. Chappie saves his creator by uploading the dying human's mind to another robot. The film leaves the outcome much more uncertain, with at least a potential for middle ground between Heaven and Hell. It is worth noting, in this vein, that Mary Shelley's 1818 novel did not see Frankenstein's monster (or creator) as inherently evil. It is thought that she was inspired by the idea of the golem.

Harry Collins and Trevor Pinch (1998) note that attitudes toward science and technology tend to be polarized between those who see them as bringing brighter futures and others who see science, along with the capitalists who commercialize its discoveries, as leading to the destruction of our world and its ecosystems, eroding what is good in our societies. Collins and Pinch argue that, rather than a savior or a vengeful monster, science is a golem. In Jewish mythology, a golem is a powerful creature made by spells that animate clay. Without careful control, it may destroy its masters with its unthinking strength. For Collins and Pinch, science and technology are not fiendish devils but bumbling giants. All tools, and the disciplines that create them, have the potential to do damage as well as useful work. The golem of science is not the problem; the failings arise from how we control its powers or how we fail to do so. If we can muddle through and get its control mostly right, we can potentially use science to prevail in the Anthropocene with all its challenges.

It is hard to deny that recent changes, as well as the ones rushing in at us, are resulting in major transformations in what it means to be human in the contemporary era. But from the posthumanist perspective that we have been presenting in this book, being human has always involved more-than-human elements. Many of these also resulted in great transformations in the genus *Homo* and the ways in which we live and lived. Before considering our contemporary intersections with machines, we examine the much longer history of our reliance on non-human forms of aid beyond our body. These have come to be known broadly as prostheses, extending the usual meaning of an artificial device to replace or augment a missing or impaired part of the body to include all devices that extend our "purely" human capabilities. A hammer, for instance, is a prosthetic extension of our force, and a megaphone an extension of the volume of our voice. For Cary Wolfe (2010: xxv), a human "is fundamentally a prosthetic creature that has coevolved with various forms of technicity and materiality, forms that are radically 'not-human' and yet have nevertheless made the human what it is." Even our greater intelligence compared to other mammals is not due to the size or sophistication of our brain; rather, it is made possible by our practices of offloading as much of our cognitive work as we can, to oral traditions, to written records, to institutions based on the division of labor, and, more recently, to computers, all of which release us from the limitations of our animal brains (Wolfe 2010: 35).

Prosthetic Extensions of Human Capabilities

These prostheses, tools of all sorts, are part of what makes humans post-human. Our capacities have been more than human for many millennia. Some will object to seeing this reliance on non-humans as supporting the claims of posthumanism, pointing out that humans are not the only species to use tools. Does that mean that apes are also post-simian? We will leave that question to primatologists, but it is important to recognize that apes' understanding of the world has presumably not been organized around simianism and ideas about simian exceptionalism (at least, other than in the fictional world of *Planet of the Apes*). Humanism generally fails to treat our non-human collaborators in the history and pre-history of *Homo sapiens* as anything other than passive resources. Posthumanists are left the task of bringing them back in and showing why it matters.

Richard Wrangham (2009: 1) argues that the transformative moment that gave rise to the genus *Homo* occurred with the control of fire and the advent of cooked meals, probably by *Homo erectus*. This occurred perhaps 1.8 million years ago, although there is considerable debate about the timing of the first human-controlled fires since, without hearths, they are hard to distinguish from natural fires. Wrangham argues that the use of fire and cooking changed "our bodies, our brains, our use of time, and our social lives." By making use of external energy, we "became organisms with a new relationship to nature, dependent on fuel" (1). In an argument parallel to the discussion of how bacteria in our guts can increase the number of calories that we can obtain from a specific quantity of food, he documents how, for many foods, cooking also makes more calories and nutrients usable by our guts. The eminent French anthropologist Claude Lévi-Strauss stated that cooking establishes the difference between animals and people and marks the transition from nature to culture. Wrangham points out that Lévi-Strauss neglected the physical consequences of this epochal shift.

Cary Wolfe (2010: 295) sees even spoken language as a prosthetic extension of human capacities. This claim is contestable, since speech is generated by our bodily organs. However, the point is much clearer for written language, which fixes in a (relatively) permanent form the immediately disappearing, except in memory, sound waves of speech. This serves to externalize our words and raises the possibility that they might outlast us and the memories of those who knew us. Jack Goody and Ian Watt (1963: 304) argue that, "in the perspective of time, man's biological evolution shades into prehistory when he becomes a language-using animal; add writing, and history proper begins." Writing "changed the whole structure of the cultural tradition" (344). Among other things, it made possible the rise of ancient bureaucratic states that required written records for their administration, particularly for taxation. From the earliest examples, texts were also used for the religious support of rulers. More controversially, Goody and Watt argue that an "easy" writing system (in terms of materials, as well as the symbols to be memorized, so that an alphabetic system is easier to learn than an ideographic system like Chinese) allowed historical inquiry, which in turn encouraged skepticism about the claims made by authorities. They argue that this allowed a critical intellectual culture to emerge in Greece, leading to democracy.

Goody later further developed these ideas on the consequences of literacy, which have become known as the "literacy thesis." They are widely discussed. The arguments for the necessity of literacy and documents for a bureaucratic state are generally accepted (Cole and Cole 2006; Sample 2015). However, the claims for causing changes in human cognition have been criticized by anthropologists. Halverson (1992) demonstrates that writing does not ensure logical thought, nor does its absence prevent it. The consequences "depend entirely on the uses to which literacy is put" so that "both the uses and the consequences are extremely variable" (Halverson 1992: 314). For the Merina of Madagascar, Bloch (1998: 161) found that literacy "has not transformed the nature of Merina knowledge—it has confirmed it," as writing developed distinctive forms consistent with oratory. In other contexts, literacy has considerable impact, particularly when imposed from outside through colonialism, but the nature of this impact varies greatly from one situation to another. This discussion raises the key issue of **technological determinism**, the idea that a technology directly leads to certain kinds of change, whether psychological, social, cultural, political, or economic (and often all of these).

Technological Determinism

For archaeologist Ian Hodder (2012: 35), various forms of media "have constituted part of human cognitive architecture since the Upper Palaeolithic. Changes in external symbol systems have altered the capacity for human memory." Even those who are most critical of the literacy thesis acknowledge that writing had the crucial effect of increasing our capacity for information beyond the individual, even when oral cultures emphasized memory skills much more than we do now (Halverson 1992: 315). Writing is, of course, only one of the ways in which the adoption of new technologies changes humans. Hodder (2012: 37) reminds us that in *The Descent of Man*, Darwin described how the "structure of the hand had evolved because of the use of tools." A post-Pleistocene (the epoch from about 2,588,000 to 11,700 years ago) "example of gene-culture co-evolution is the spread of lactose tolerance linked to the spread of domestic cattle and milk." For Hodder, the physical human body is as it is "because of its interaction with things. And the same can be said of human cognition" (37).

It is not helpful to interpret as technological determinism the interactions between humans and the non-human things that they rely on. Hodder believes it is more useful to think in terms of what things and tools make possible, which is usually both diverse and limited. Similarly, in a study of the introduction of new information technologies into corporations, David Hakken (1993) concludes that the technologies themselves make many different kinds of organizational change possible. The actual configurations and uses adopted are not determined by technology but rather through technology's interaction with the agendas of powerful actors, the structures of the economy, and the potential for resistance by employees.

While rejecting technological determinism, Hodder (2012) also rejects the assumption that it is only human intentions and their social and cultural organization that influence whether and how "things" might affect us. He argues that our shared "connections and dependences have implications and consequences of which people are not aware and which may be unintended" (103). His method involves taking things "in themselves" seriously, attending to "the physical, chemical, engineering dimensions" of objects rather than always embedding them in meaning and discourse as material culture studies tends to. There are similarities with ANT here, respecting non-humans as actants. Hodder suggests, however, that ANT gives insufficient attention to the ways in which humans and things "entrap each other" (93). As we have seen in Chapter 2, Latour (2005) rejects the human–non-human dualism, but Hodder argues that doing so risks "losing one of the main motors of change—the limited, unfixed nature of things in themselves and their relationships with each other" (Hodder 2012: 93), not necessarily mediated through humans. Things like the Balinese roosters discussed in Chapter 3 are more than what they mean to people. Non-human entities, whether animal or inanimate, have their own dynamics, beyond their involvement with humans, which are influenced by their material characteristics, such as the rusting of iron, and the influence of the environment, such as how wet it is. Because of these changes that happen independent of our intentions (i.e., they are unfixed), we have to bring things into consideration in our analyses as more than passive resources used for human purposes.

A useful term that allows us to simultaneously recognize that things such as a hammer make certain things easier (pounding nails) than others (opening your smartphone to repair it), while avoiding technological determinism, is **affordance**. Introduced by James Gibson (1979) in his influential work on organisms' perception of their immediate environment, he defined the term as the potentialities of an object for a particular set of actions. Fire can afford making certain foods edible or more nutritious, but it can also result in injurious burns (Hodder 2012: 49). An Internet-connected smartphone may increase your productivity by providing faster access to essential information, but it may also reduce it by encouraging you to procrastinate by playing *Angry Birds* or other games. For archaeologists, thinking in terms of affordance is particularly useful because they often have to interpret an artifact based on its shape and characteristics (e.g., does it hold liquids inside or do they seep through quickly?) alone, without information on how people actually used it beyond clues from patterns of wear and so on.

Humans are deeply dependent on things and have become progressively more so over time as the average household contains more and more "stuff." Hodder (2014: 20) distinguishes between "dependence," our reliance on things, and more pathological relations of "dependency" that limit our "abilities to develop, as societies or as individuals." For example, our economy is dependent on high rates of consumption that are causing increasing problems of sustainability. Our global information and communications systems use as much electricity as Japan and Germany combined, for example (Hodder 2014: 27). The dependency that we develop on things (and vice versa) produces a "stickiness," which Hodder argues is "inadequately dealt with in the networks and heterogeneous hybrids of ANT." Networks for ANT are always subject to change as new actants move into place, but Hodder thinks that the openness in this approach pays insufficient attention to the way in which we tend to become "entrapped" in our dependencies with certain things.

There are some similarities in Hodder's ideas to the more common idea of path dependency. **Path dependency** examines how investments of effort and resources in our past choices tend to keep us following that same developmental path. Past actions become sedimentary layers of contemporary landscapes, political institutions, and social structures. While the past does not determine the future, it makes particular directions easier to pursue through conditioning human action. Path dependency helps explain the tendency of complex systems to produce unintended consequences that frustrate the best efforts of managers and planners (Ghitter and Smart 2008). The classic example is our continuing use of the standard QWERTY keyboard, which was originally designed precisely to make it hard to type quickly, to prevent the mechanical typewriter keys from tangling. Most of us continue to use it despite the availability of more efficient designs because of the sunk costs invested in learning the QWERTY layout and due to its presence all around us. Hodder rejects path dependency inasmuch as it tends toward determinism. He believes that we can also contribute to forging our paths. While we do not always continue to follow the easiest course, greater effort is required to shift from the beaten path. It has been persuasively argued that shaping new paths requires understanding the tendencies toward particular paths to "keep the path from kicking back in" (Torfing 1999). Such openness is afforded by our entanglements with things, which "are always provisional, worked out in practice, temporary and partial. Things are always happening that have to be dealt with in a continual boot-strapping process. Simple material events in one domain might have all sorts of unexpected spin-offs" (Hodder 2012: 110). Because of this instability, people constantly chase along chains of dependence to fix things. Doing so draws them into further dependences. Things fall apart, like houses, and have to be fixed. Fixing them, though, usually traps us in further commitments, as with the continual patches to buggy computer programs.

Media and Sociality at a Distance

Another key dimension of the human prosthetic past is the way that media facilitate communication at a distance. They allow us to speak, hear, and see things that happen beyond our immediate reach. As a discipline, anthropology has until recently emphasized lived experience and doing research by staying in particular places for long enough to write an ethnography of a way of life. Increasingly, however, our lives are experienced to a much greater extent through the mass media. Arjun Appadurai (1996) sees a key feature of globalization as the possibility that we can choose lives based not on relationships and experiences in our own locality but on possible lives offered by mass media, in recent decades supplemented by social media. A 2010 survey in the United Kingdom found that when asked the question, "What would you like to do for your career?" 54 per cent of teenagers answered, "Become a celebrity."

Such developments are not completely new. Printing presses helped to disseminate dissenting views through pamphlets and books, encouraging Catholics to reimagine themselves as Protestants who aspired to a direct relationship with God rather than one mediated through the priestly hierarchy. Enlightenment ideas emphasized questioning taken-for-granted ways of life. In addition to books, the key Enlightenment scholars collaborated on a great encyclopedia to encourage popular education. In the political realm, the Enlighteners asked why authorities had power and proposed better ways to arrange society. Benedict Anderson (2006) described the emergence of nationalism in colonies as an "imagined community" made possible to a large extent by new mass-circulation newspapers that helped people to think of themselves as a nation rather than a collection of precolonial localities or provinces of empires. However, such imagined communities are not limited to nationalism. The invented language Esperanto was created by Ludwig Zamenhof in the 1870s to foster harmony between people from different countries, believing that separate languages contributed to warfare. At present, the threats of climate change and pollution are encouraging many to think of themselves as global citizens as well as, or instead of, citizens of nation-states.

According to Stuart Hall (1977: 340), the mass media have "colonized the cultural and ideological sphere. As social groups live ... increasingly fragmented and sectionally differentiated lives, the mass media are more and more responsible for providing the basis on which groups construct an 'image' of the lives, meanings, practices and values of other groups and classes." In other words, more-than-human technologies provide the basis on which humans come to understand themselves, a profoundly posthuman situation. Most new media are less centralized but also affect the condition of being human in the contemporary world. Social media have been widely adopted by repressed dissidents in different countries, even against strong governmental interference, for example. The easy portability of cassette tapes contributed to the 1979 Iranian revolution by allowing the recorded speeches of Ayatollah Khomeini to be smuggled in and played surreptitiously in mosques. Fax machines were key to maintaining communications between activists and global media in China's Tiananmen Square protests in 1989, when the government cut other lines of communications. Photocopiers, and their predecessor, the mimeograph, helped the circulation of underground publications (*samizdat*) in the Soviet Union. All of these technologies were more accessible than the mass media of television, radio, and newspapers, controlled by state organizations in these countries. More recently, Twitter and Facebook helped circulate information, including about the organization of protests, during the Arab Spring.

Declining **economies of scale** for media may have the effect of demassing them, at least partially. Print, film, and television usually must produce many copies to be profitable. New technologies can manage shorter runs. This happened early on with photocopiers, cassettes, and faxes, later with email lists, the Internet, Napster, and blogs. Digital distribution is nearly cost-free, with the result that it can threaten existing companies and commodity chains. However, new media may also help maintain traditional activities and values. Broadcast media and the Internet may be a better match for smaller language groups than the print media, which could contribute to cultural survival. Consequences depend on the nature of the media. A variety of alternative paths that might be pursued are often equally facilitated by new media, heightening the importance of cultural, economic, and political factors that influence who has the ability to mobilize such media to further their interests and goals.

Transhumanists see new media like the Web and virtual reality as contributing to the radical transformation of humanity, rather than as being channeled by pre-existing structures. Current and emerging media blur lines between self and technology. GPS, smartphones, and tablets allow us to externalize our memory. Studies have found that GPS usage affects our cognitive maps, with a negative correlation found between GPS usage and drawing maps on a city scale (Minaei 2014). There has been rapid growth in self-monitoring, through Fitbit and other technologies. See, for example, http://quantifiedself.com/ with the slogan "self knowledge through numbers." People are beginning to archive their whole lives by always being on cameras.

Yet how much has really changed? Denise Carter (2005) provides ethnographic evidence that friendships in cyberspace are not fundamentally different from those in "real life." Radical possibilities do not always result in radically new social forms. Tufekci (2008: 544) found that the best predictor of whether students at a US college adopted Facebook was their attitude to social grooming. Non-users display an attitude "towards social grooming (gossip, small-talk and generalized, non-functional people-curiosity) that ranges from incredulous to hostile." Non-users have as many close friends as users, but keep in touch with fewer people. Women are more likely than men to be users. This suggests that people are using social media in ways that are consistent with their prior interactional preferences. If so, the transformations may be less than often assumed. When we ask our classes what they use social media for, it is common to hear that it is mostly for keeping up with friends and family, so that it reinforces existing bonds rather than forging many new ones at a distance. The experience of migrants is different than for those living where they have grown up, but their telecommunication use also preserves relationships back home. Fast Internet and Skype make much richer distance communication possible with loved ones and friends left behind. Research generally finds that cheap cellphones and digital money transfer systems like M-PESA in the global South have very high returns on investment for the poor. South Asian fishers and farmers who used to have to take whatever price the local brokers offered, in part because they had no access to information about current market prices, now can check beforehand and have a better negotiating position. Yet, in the case of middle-class North American students, we are dubious about the economic returns on their telecommunications investments. Those in need seem more likely to make sure they get the most advantage out of their investments. For the more affluent, powerful handheld technology may only provide more convenient access to information that they could already access through their computers or libraries (if they so wished), making them redundant channels of information in contrast to the novel vistas for new adopters of information technology in the global South.

As Nancy Foner (2000) convincingly demonstrated in her comparison of transnational relations of migrants to New York in 1900 and 2000, the frequency of communication does not determine its qualitative importance. Despite the great increase in the speed and availability of communications and transportation technologies, and the decrease in their costs, she found that the ties between migrants and their homelands were at least as strong in 1900 as in 2000. Phoning every day does not necessarily have greater qualitative importance than occasional, carefully crafted letters. Again, we must beware the temptation of technological determinism. Media and technology enable, but in doing so they offer multiple paths and possibilities, the adoption of which is affected both by needs and by preferences. Often, once we move beyond the hype of both optimists and pessimists, continuities of cultural patterns are more apparent than radical departures, as new technologies are adopted to fit with existing ways of doing things. Still, in the longer run, major changes can emerge even while initial outcomes are kept comfortably within the constraints of current institutions. Our research among migrants from rural areas to coastal cities in China found that the migrants often bought mobile phones for their parents to make it possible to stay in touch. This often brought older people into contact with modern communication technology for the first time. Many subsequently used it for diverse other reasons.

The last century has indeed seen an accelerating set of waves of new technology that have created or intensified connections between the people of the world, even those who were quite isolated before World War II. Anthropologists came to pay considerable attention to the impact of mass media on previously non-literate populations (Boyer 2012; Michaels 1987). Much of this work rejects ideas of the wholesale displacement of local difference, and instead stresses that what is most successful is often due to its compatibility with cultural preferences. Brian Larkin (2008) found, for example, that Bollywood films were more popular than Hollywood films in northern Nigeria because of the familiarity of the concerns about family involvement in marriage choice and the struggle to find ways to integrate love with family obligations. The individualism of Western films was much more alien, beyond the attractions of special effects.

More recently, attention has turned to the impact of computers, cyberspace, and other digital technologies. Almost all of the initial research was done in and about the West (Hakken 2004). An early exception was Arturo Escobar (1994: 138), who asked if there were possibilities for Third World societies to "participate in cyberculture without fully submitting to the rules of the game." Researchers were encouraged to conduct basic research on non-Western computing, with particular attention to distinctive forms of adoption, interpretation, and innovation. A good example is what has come to be described as India's "frugal innovation," which has come to receive considerable attention in Western business schools. David Hakken (2004: xi) argues that doing research on non-Western use of computing and cyberspace not only expands our knowledge of how they are incorporated into particular cultural contexts, but also provides opportunities for testing the extent to which the technology is culturally mediated and transformed. The editors of a volume on digital anthropology conclude that research so far confirms our discipline's past observations of humanity's remarkable ability to "reimpose normativity just as quickly as digital technologies create conditions for change" (Miller and Horst 2012: 4). Digital technology is not necessarily producing greater similarities among the world's people, but is often implemented in ways that maintain existing systems of hierarchy or bureaucratic procedure. Technology-inspired hype about how "this changes everything" has generally failed to represent the complex blends of innovation and conservation of past practices that are more likely to result. Assumptions about the consequences of a digital or IT or computer revolution can be dangerous when they are incorporated into important economic policies and social programs (Hakken 2003: 4). As these studies argue, we cannot usefully or safely make assumptions about the impact of technologies without carrying out careful ethnographic research into how digital technologies are actually used by people. We cannot just accept the predictions of their promoters (or opponents).

Cyborgs and Transhumanism

In the rest of this chapter, we turn to discussions of the future, with an emphasis on cyborgs, **distributed cognition**, and transhumanism. **Cyborg anthropology** originated as an interest group within the American Anthropological Association's annual meeting in 1993, after the first panels on the topic in 1992 (Downey, Dumit, and Williams 1995). Donna Haraway's 1985 article "Manifesto for Cyborgs" (republished in 1991 as *A Cyborg Manifesto*) could be considered the founding document of cyborg anthropology. In one sense, cyborg anthropology is the empirical study of the incorporation of technology within human bodies and the addition of prosthetic extensions to those bodies. It examines what this means for humans and our future. However, Haraway's "Manifesto" was opposed to the militarization of cyborgs, well before the emergence of the "soldier of the future" programs (mentioned in Chapter 1). The word "cyborg" was first used in ideas about adapting humans to space travel in 1960 (Garreau 2006). In Neal Stephenson's 2015 novel *Seveneves*, gene editing is used to create seven different genetically distinct races in orbital habitats when the earth becomes uninhabitable after the moon is shattered. We return to recent major advances in gene editing below.

Haraway (1991: 151) noted that, despite the cyborg's usefulness in blurring the boundaries between human/animal and machine, the main trouble with cyborgs is that "they are the illegitimate offspring of militarism and patriarchal capitalism." However, she hoped that in this case illegitimate offspring would be "exceedingly unfaithful to their origins." Her essay was a provocation to feminists "who wanted to position women in alliance with nature and against technology" (Hayles 2006: 159). Haraway (2003: 4) says, "I published 'The Cyborg Manifesto' to try to make feminist sense of the implosions of contemporary life in technoscience." She rejected feminist approaches that positioned women with nature and opposed them to masculinist technology. Haraway felt that a forward-thinking feminism needed to engage with, rather than reject, technoscience. By drawing out the many implications of the blurring of boundaries between humans and machines, she stimulated an avalanche of important research. Not all of it followed her own idea that cyborgs had the potential for providing a new basis for progressive politics. Some of her own illegitimate offspring have returned to a much less ambiguous promotion of cyborgs in the form of the transhumanist movement.

A considerable amount of cyborg anthropology research has focused on surgical implantations, originally rather crude and without significant informational components, such as pacemakers, but more recently involving much more fundamental linkages between the human and the machine, such as brain–computer interfaces. Another relevant example is cochlear implants, hearing aids that not only amplify sound but actually provide a connection with auditory cells. Cutting-edge interventions often focus on people with disabilities, as therapeutic interventions are more acceptable ethically, particularly for life-threatening conditions (Garreau 2006). With rapidly developing prosthetic technologies, however, the questions are changing in fascinating ways. For example, Oscar Pistorius was initially ruled ineligible to compete in the 2008 Olympics. His blade-enhanced legs allowed him to use 25 per cent less energy than able-bodied athletes to run at the same speed. Is he "disabled, or too-abled?" In 2008, *Time* magazine selected him as one of the 100 most influential people, citing him as "on the cusp of a paradigm shift in which disability becomes ability, disadvantage becomes advantage" (quoted in Camporesi 2008: 639). Subsequently, design rules have been put in place that attempt to prevent the possibility of runners using blades breaking non-Paralympic world records. As the controversy over the omission of the entire Russian team from the 2016 Olympics for chemical doping demonstrated, athletic performance is a realm in which the boundaries of nature are policed, albeit in the pursuit of *almost* (but not quite) superhuman achievements (Gibson 2015). Should the Paralympics opt not to limit technological enhancements in the future, it is entirely likely that in many categories, such as swimming as well as running, the Paralympic records will surpass those of the Olympics. Is our definition of athletic excellence limited by the maximum capabilities of species-typical individuals, or will an unlimited Paralympics become more like motor sports? Still, many devices, like poles for vaulting, ice skates, and bicycles, are already expanding human capabilities dramatically.

Research on disabilities and impairment has been leading the way for an understanding of the emergence of more-than-human capabilities, with new prosthetic technologies increasing abilities beyond the species typical (Gibson 2015). In the 2013 documentary *Fixed: The Science/Fiction of Human Enhancement*, a rock climber who lost the lower part of both legs felt that he had become an even better climber after he developed a set of specialized prosthetics adapted to terrain challenges, and which could be swapped mid-climb.

There is a wide literature on the anthropology of disability that traces the cultural and social implications of shifting ideas within biomedicine, but which we can only briefly touch on here. Two anthropologists who have contributed greatly to the field, Faye Ginsburg and Rayna Rapp (2013), have described their work as "entangled ethnography," where the researcher has a stake in the process being documented, because it began with their own experience as parents of children with learning disabilities. Their distinctive roles led them into research laboratories, educational bureaucracies, and activism. It was also entangled in the sense used in this book because the topic led them to be "promiscuous violators of the walls erected by medical manuals and school bureaucracies" as they attempted to understand "how disability takes shape in concrete, cultural locations" (Ginsburg and Rapp 2013: 192).

Most recently, following the tangles where they have led has brought Rayna Rapp (2015) into studying research on learning disabilities incorporating big data into new models of the brain. One of the current big science projects, like the Human Microbiome Project discussed in Chapter 2, is the effort to understand the connectome. The connectome is the total set of neuron connections in the brain. A key part of the research involves an attempt to map the connections that are made in the brain under different conditions of activity through functional MRI scans. The increased capacity to analyze big data and mine it for unexpected connections is a major element of the connectome project, and its importance for the study of disabilities can be seen in its potential to discover overlapping "circuits" and "hubs" that are implicated in multiple disabilities, a better understanding of which might eventually improve diagnostics and treatment. In many ways, this kind of big neuroscience takes things away from the everyday concerns of parents of children with disabilities, but Rapp notes two unintended positive consequences. First, the great sophistication of the analysis makes it possible to undermine the dualistic distinction between "healthy" and "disability," which has increasingly been replaced by "neurotypical" and "neurodiverse," and recognizes that there is a spectrum, rather than a single "healthy" type. Second, this more nuanced discussion is having some effect in encouraging the New York Department of Education to integrate ideas about neurodiversity into their educational programs. At the intersection where the "research claims of neuroscience big data and small kids bearing diagnoses meet, more capacious understandings of the range of human variability may emerge" (Rapp 2015: 18).

New Reproductive Technologies

Another field in which technological intervention has been controversial, and which possesses the capacity to transform fundamental questions about humanity, nature, and culture, is the topic of technologically assisted reproduction, particularly invitro fertilization (IVF). This field is connected in various ways with the anthropology of disability. Rayna Rapp (2004: 3) found that amniocentesis was turning women into "moral pioneers," situated on a "research frontier of the expanding capacity for prenatal genetic diagnosis" and "forced to judge the quality of their own fetuses," making "decisions about the standards for entry into the human community." Some authors are talking about the children created by the recent intensification of their technologically mediated conception, birth, and rearing as producing a generation of "cyborg babies," often in pursuit of perfect babies capable of competing in tough labor markets (Dumit and Davis-Floyd 1998). New "technosocial practices have troubled boundaries between nature and culture, matter and spirit, love and money, life and death, and individual and collective" (Roberts 2012: xxii). Yet, the impact is not determined by the technology itself, but rather depends on the social and cultural context within which these new technologies are deployed. How societies, and groups within them, accommodate technological possibilities that hold the potential for disrupting cherished institutions and values varies greatly.

When a group of anthropologists became involved in research and policy recommendations on assisted reproduction, including not only IVF but also surrogacy, they struggled with the question of framing it within, or without, questions of family values. They resorted to emphasis on kinship rather than family. This enabled them to address issues of relatedness without assuming particular kinds of families (Edwards et al. 1999). Since then, there has been a rich fluorescence of anthropological research in this field, particularly from people working from the perspective of feminist science studies. As such authors stress, research on new reproductive technologies shifts the study of kinship from parents, children, and other relatives to a much wider arena of people related in various ways, including medical professionals and others with technical expertise, revitalizing an anthropological perspective on kinship that never confined its significance within a narrow idea of "family."

Elizabeth Roberts (2012) studied IVF in Ecuador and found that it was more readily accepted there than in the United States. She argues that this is because in Ecuador, unlike in the United States, nature is not seen as a fixed object to be discovered by people. Instead, it is experienced as shaped through interactions with people who in turn exist in relation to the biological world—other people and divinities in situations of interdependence, rather than individual autonomy. Images of the Madonna and prayers by medical staff were common in IVF clinics, for example. Other anthropologists, however, question whether IVF does indeed trouble Western ideas of nature and culture. Sarah Franklin (2013) argues that IVF was quickly normalized because "it already belongs to techniques of normalization," not only including marriage, kinship, and gender, but also scientific progress, livestock breeding, baby showers, consumer culture, and medical technology. The normalization of IVF is a kind of "hybrid culturing" that allows new technology to coevolve with existing social institutions, particularly prevailing gender and kinship norms (Franklin 2013). Yet, even when IVF is channeled to fit with pre-existing conceptions of nature and social institutions, the specific material character of the reproductive technologies can cause more subtle transformations. These often arise from individuals undergoing IVF who experience a profound technological ambivalence, resulting from complex and difficult interventions in a procedure that has come to be seen simply as medical assistance for natural desires to have a family (Franklin 2013).

Smart Homes and Cyborg Cities

Shelter is another domain in which technologies have begun transforming practices essential to humanity since our species emerged. Our dwellings are increasingly not only places of privacy, safety, and meaningful sharing, but also places where we are more intensely connected to the wider world than anywhere else (if only because of cheap Wi-Fi), except for those whose work requires digital interaction. Smart homes promise to take this connectedness further and to make the home itself an information-rich extension of our bodies and desires. Individual homes are only components of a broader process, known as smart city strategies, of embedding increasingly complex networks of sensors and automation throughout cities. Some scholars have begun talking about "cyborg cities." They see urban areas as a kind of fusion between human and non-human components, dependent on an urban metabolism in which, to survive, the city draws on the flow of materials, water, air, and power from vast hinterlands (Gandy 2005). Making homes smarter is often justified by the claim that doing so can help to make them more energy efficient and contribute to reducing our carbon emissions (Pink et al. 2016).

Spending on smart-home appliances is expected to grow from $40 million US in 2012 to $26 billion US in 2019. Research has also been growing on the topic, but the majority of this has been carried out by engineers. Only a small percentage of publications are based on research with users in actual smart-home environments (Wilson et al. 2015). Anyone—at least anyone of our generation—who has tried to program a smart remote control will realize that engineers commonly fail to produce intuitive interfaces, and this has been a major failing of most smart-home systems so far (Leitner 2015), forcing us to adapt to them rather than vice versa (Lanier 2010). New technologies are rarely used as their designers intended because they "enter pre-existing environments that are contested, emotionally charged and dynamic." Without research into how actual households currently manage their complex and challenging lives, designers are unlikely to recognize—and may complicate rather than complement—the ways in which "apparently chaotic domestic environments" possess their own "smartness," in the way, for example, that households manage communication and organize the flow of clutter and mess through the home (Wilson et al. 2015: 470). Once we incorporate the diversity of kinds of household in a single city, and start to envisage the differences between societies, knowing how to automate everyday activities seems a daunting task, unless it is done in ways that build upon the skills that households already deploy to cope (Pink et al. 2016). Companies are committing large investments to smart homes, with Google alone spending $3.2 billion to acquire Nest, which makes smart thermostats, and $550 million to buy Dropcam, which makes home-security cameras, but sales have been disappointing. Only 6 per cent of US households have a smart-home device ("Where the Smart Is: Connected Homes Will Take Longer to Materialise than Expected" 2016).

One of the few studies of smart homes by an anthropologist notes that there has been little critical investigation of the effects of smart-home technologies on everyday lives (Strengers 2016). Studies of the "homes of tomorrow" from the early twentieth century reveal that utopian visions for our future have unfolded very differently from how they were imagined. The connection of the home to the wider world is clearly not a completely new thing. In Western cities, a great transformation of houses took place from the 1920s to the 1950s, when "domestic space was designed around a technical core of sewers, water and gas mains, power cables, and telephone lines" (Putnam 2006: 145). These infrastructures changed the home into a hub of wider systems, which are now taken for granted as just part of ordinary houses, even when the majority of the world's dwellings are still at best partially connected (e.g., they may have cellphones and illegally pirated electricity, but not piped water or sewer connections).

Current smart homes are similarly characterized by utopian aspirations alongside a "disturbing absence of social research conducted with people who actually live in them" (Strengers 2016: 62). It is not clear that real demand exists—yet, at least—for smart homes; rather, the pressure so far has been from suppliers and utilities (Wilson et al. 2015), often encouraged by government subsidies for items like smart thermostats to save energy. Proponents of the smart home are trying to intervene in everyday life to manage energy in very specific and quantified ways (Strengers 2016).

As always with ethnographic approaches, there are surprises. One group of early adopters has been Orthodox Jews, "who have been using home automation technology for decades to coordinate the religious practice of resting on the Sabbath, and to maintain the modern interpretation, which is that it is forbidden to turn electrical devices on or off on this day" (Strengers 2016: 70). Lighting and appliances that might have stayed on, or off, for the full Sabbath are scheduled to operate at particular times, so that there is no need to operate machinery to have its conveniences, or to waste power by leaving the lights on all day. Note that, depending on the particular practices adopted, this might have the effect of either reducing or increasing energy use. In a Halifax synagogue, there has been disagreement about a plan to have the elevator operate continuously on the Sabbath, stopping at each floor in succession. Once again, we see that technological determinism is not the best way to think about the potential impact of emerging technologies; in most cases, if technologies are adopted, they will be fitted into existing routines, although in the longer term they can produce significant changes as well (Franklin 2013).

Cyborg Animals

It is not only humans that are becoming cyborgs. Robinson et al. (2014) are working on a design for an alarm system that would allow diabetes alert dogs to remotely call for help when their human falls unconscious due to a hypoglycemic attack. They report that they are using a "multispecies ethnographic approach" to discover the requirements for a "physical canine user interface, involving dogs, their handlers and specialist dog trainers." They found tensions between the requirements for canines and the human users, and call for "increased sensitivity towards the needs of individual dogs that goes beyond breed specific physical characteristics" to move from "designing *for* dogs to designing *with* dogs." This is only one example of a growing field of animal–computer interface. Governmental efforts to stop the spread of BSE, discussed in Chapter 2, resulted in Canada, as well as other countries, requiring the attachment of radio frequency identification device (RFID) ear-tags which can be read with hand scanners at auctions and slaughterhouses to verify the age of the animal. This also allows authorities to track the animal back through its life to identify other animals that might have been exposed to prions. The design of these ear-tags could have benefited from some multispecies research, or even just old-fashioned talking to the farmers, however. We were repeatedly told that in local conditions, the tags were very often ripped out after becoming snagged in vegetation. Dealing with lost tags was difficult, with the result that, rather than attaching the tags to calves, as required, farmers would usually just keep the record number and attach the tag before selling the animal, largely defeating the purpose of the legislation.

The **informationalization** of livestock and crops has become a growing trend throughout the agri-food industry, partially to be able to meet regulatory requirements, but also to satisfy consumer demand for local, organic, or ethically raised food. Drones and GPS-controlled cultivators are also facilitating the emergence of "precision agriculture." Researchers are developing sensors that could be implanted inside cattle to identify digestive problems, the early stages of an animal going lame, or the peak period of estrus for insemination ("Stock Answers" 2016). These interventions have the (probably) unintended consequence of making it much harder for small producers to compete with larger producers who can more easily afford the accounting and equipment costs, and can benefit from the information that becomes available about market conditions (Smart and Smart 2012b).

The Cognisphere

Informational extensions of our capabilities have also been a key area for cyborg studies. As Katherine Hayles (1999) argues in *How We Became Posthuman*, a key dimension of the idea of the cyborg involves

> informational pathways connecting the organic body to its prosthetic extensions. This presumes a conception of information as a (disembodied) entity that can flow between carbon-based organic components and silicon-based electronic components to make protein and silicon operate as a single system. When information loses its body, equating humans and computers is especially easy, for the materiality in which the thinking mind is instantiated appears incidental to its essential nature. (2)

Hayles is fundamentally critical of the idea that we can equate human intelligence with disembodied informational patterns, which theoretically could allow for the uploading of consciousness (or *Star Trek*–style transporters). She insists that our minds are thoroughly embodied and inseparable from that bodily context.

Despite her important work on the subject, Hayles (2006: 159) has more recently argued that the cyborg has become a less fruitful metaphor to work with because "it is not *networked* enough" (emphasis in original). Although they continue to become more sophisticated, bodily implants are less central to current trends than "distributed cultural cognitions embodied both in people and their technologies" (160). If brain chips making direct connections between brains and computers come into widespread use, the cyborg may eventually have its day. For now, Hayles feels that the cognisphere, rather than the cyborg, is "the compelling metaphor through which to understand our contemporary situation." The cognisphere, the globally interconnected cognitive systems of which humans are only one part, "is not binary but multiple, not a split creature but a co-evolving and densely connected complex system" (165). Most Internet traffic no longer involves people, but is computer to computer. This trend is growing faster with what has come to be called the "Internet of things," direct linkages between gadgets that used to be unconnected and without information-processing power, such as toasters and washing machines. This concept is generating manufacturer visions of refrigerators that can order milk from online stores when it is running low, or bridges with embedded sensors that can signal their imminent collapse. Cognition is in the "cloud," in the earth, and under the seas, distributed through complex networks of heterogeneous actants. It enables us to do many things we couldn't in the past, but its distributed workings transcend our understanding. It is perhaps this inscrutability that gives us the impression of information having a life of its own. Its ecology grows constantly, and corporate data-mining follows. Some say we are now in a new age of **big data** and analytics that glean useful information in the pursuit of profits. Others worry that artificial intelligences designed to find patterns and profits in the stock markets may eventually not only cause financial crashes but could gain access to sufficient information and processing power to achieve the Singularity. **Artificial intelligence** is now rivalled by the field labeled "**artificial life**," which uses evolutionary principles to generate the survival of the fittest software programs, rather than relying on human designers working directly on the code (Helmreich 1998).

Gene Editing and Transhuman Futures

Merging with machine intelligences is only one path to a transhuman future; altering the human genome is another. The ability to manipulate genes has recently taken a large step forward with the development of a generalized tool for gene editing, CRISPR (clustered regularly interspaced short palindromic repeats). The August 22, 2015, cover of *The Economist* (an influential business weekly) featured the title "Editing Humanity: The Prospect of Genetic Enhancement." Editing genes has been taking place for decades, but most techniques were specific to a particular species or gene, whereas CRISPR, derived from a species of bacterium, is a generic tool simple enough to be used by a high school student (or, worryingly, terrorists without access to a cutting-edge laboratory). One major risk is the ability to create organisms that incorporate CRISPR into their own genome so that they can edit their own genes and "spread themselves through a population with blithe disregard for the constraints of natural selection" ("Genome Editing: The Age of the Red Pen" 2015: 20). Such modifications could have benefits, such as making mosquito vectors of malaria unable to carry the disease, but the dangers are also momentous. Attention to this new tool is intense: three of the ten most-cited scientific papers published in 2015 concerned CRISPR. The editing of human germline (sperm and egg cells) genomes that could be passed to the next generation is particularly controversial but has the potential to end genetically inherited diseases such as Huntington's. Genetic modification for enhancement rather than the prevention of genetic flaws is even more contentious. Many scientists are calling for a moratorium on all germline modification experimentation: research on human germlines is currently banned in 40 countries. Such voluntary restraint may be seen as supporting Garreau's Prevail scenario. An international summit in Washington, DC, organized in December 2015 by the national academies of the United States, the United Kingdom, and China, failed to agree on a widely sought ban on gene editing. According to the statement adopted, "It would be irresponsible to proceed with any clinical use of germline editing unless and until the relevant safety and efficacy issues have been resolved, based on appropriate understanding and balancing of risks, potential benefits, and alternatives, and there is broad societal consensus about the appropriateness of the proposed application" (Sample 2015).

Ethical Debates about Transhumanism

Similar accounts of the potential for currently emerging technology, such as nanotechnology, to facilitate the transition to transhumanism could be elaborated if space permitted. But enough has been said, perhaps, to make it clear that the transhumanist dream is in the process of moving from futurist dream to current potentiality. With this emphasis on progress, transhumanism—unlike the version of posthumanism we have been presenting in this book—clearly shares humanist ideals of progress. Nick Bostrom (2005) argues that transhumanism has its roots in rational humanism with its ideals of human perfectibility, rationality, and agency. He finds that most cultural conservatives have "gravitated towards transhumanism's opposite, bioconservatism, which opposes the use of technology to expand human capacities or to modify aspects of our biological nature. People drawn to bioconservatism come from groups that traditionally have had little in common. Right-wing religious conservatives and left-wing environmentalists and anti-globalists have found common causes, for example in their opposition to the genetic modification of humans" (Bostrom 2005: 18). For example, in 2000, George W. Bush established the President's Council on Bioethics, which released multiple reports that condemned the use of new biotechnologies to alter human minds or bodies. Their 2004 report argued that genetic and reproductive technologies undermine the value of life and threaten the natural relationships between parents and children. Attempting to improve on what we now have is seen by the report's authors as hubris and a threat to human dignity (Naam 2005). Such hubris is certainly common among proponents of transhumanism, who argue that we should "cast aside cowardice and seize the torch of Prometheus with both hands" (Simon Young, quoted in Peters 2015: 134). A more down-to-earth argument is that the search to enhance ourselves is itself a natural part of being human, that there is no clear boundary between healing and enhancement. Trying to enforce a sharp boundary between prohibited enhancement and natural healing, then, would cause massive amounts of suffering to continue. People should have the right to do what they want with their bodies, goes this line of thought. A ban on enhancement wouldn't work, but would create black markets and incentives for research in rogue states (Naam 2005). As Katherine Hepburn said in the 1951 film *The African Queen*, nature "is what we are put in this world to rise above."

Films like *Transcendence* see even good intentions as having horrendous implications since absolute power corrupts absolutely. Others assume that the expansion of capabilities will be used by many, such as terrorists, with nefarious purposes. One contemporary Hell scenario involves the radical transformation of soldiers in an arms race. Considerable research is already under way. The US Defense Advanced Research Projects Agency, renowned as the inventor of the Internet, is now in "the business of creating better humans." The head of the future soldier program claimed that "soldiers having no physical, physiological, or cognitive limitations will be key to survival and operational dominance in the future ... imagine if soldiers could communicate by thought alone.... Imagine the threat of biological attack being inconsequential. And contemplate, for a moment, a world in which learning is as easy as eating, and the replacement of damaged body parts as convenient as a fast-food drive-through" (Garreau 2006: 22). Or, for its critics, imagine soldiers whose body chemistry can be adjusted at a distance to overcome their fatigue, fear, and ethical reservations, making them efficient killing machines, constantly upgraded to keep ahead of the enemy.

Many other fears and concerns proliferate in discussions about the impact of technology already being deployed. In *You Are Not a Gadget*, Jaron Lanier (2010) worries that all the power and money generated online has begun to accumulate around those who control highly secretive systems, many of which are spying operations designed to gain our personal information. Along with many others, he is deeply concerned that the result will be the disappearance of most middle-class jobs, facilitated not only by outsourcing to other countries but also by the ability of robots and self-driving vehicles to replace many—perhaps eventually most—human workers. An unhealthy divide may arise between the technocratic elite who can analyze and profit from the big data flows, and a growing body of precariously employed or unemployed people who do not have the skills for the jobs that can't (yet) be automated. Other books with titles like *The Coming Jobs War*, *The Rise of the Robots*, and *Humans Need Not Apply* express the pessimism about the coming obsolescence of most human workers.

In the Prevail scenario, the future is not predetermined; it is "full of hiccups and reverses and loops, all of which are the product of human beings coming to grips with their own destinies. In this world, our values can and do shape our future. We do have choices" and are not merely at the mercy of large forces (Garreau 2006: 12). Garreau's (2006) main proponent of the Prevail scenario is Jaron Lanier, for whom our future trajectory is measured by its impact on human society. Garreau argues that "even if technology is on a curve, its impact is not. This is why [Lanier] is skeptical about the idea of a Singularity" (196). Doubt about the inevitability of technology-driven futures is common to both Heaven and Hell, but Prevail "is based on a hunch that you can count on humans to throw The Curve a curve" (196). Lanier (2010) assumes that a machine that surpassed human intelligence would probably be conscious only for a few nanoseconds before a software bug crashed it down. One needs only to reflect on the ubiquitous bugginess of even the most advanced computer programs to have doubts about the inevitability of the path to transcendence.

The future Lanier (2010) thinks is worth trying to bring about is one measured by an increased human connectedness that enables a myriad of interesting new directions, rather than the unadventurous ideas of a single straight-line development. His bigger worry is that software designers are forcing us to rework ourselves into dehumanized versions that match the (low) expectations of their programs. He demonstrates that poor design choices become "locked in" (another way of discussing path dependency) and difficult to change in future projects. He worries that the "new designs on the verge of being locked in, the web 2.0 designs, actively demand that people define themselves downward" (Lanier 2010: 19). To prevail, we need to reject the corporate homogenization of the tools with which we connect, and find better ones that will allow many thousands of strange new ways to reach out and become involved in making many new futures, not just a single one. This vision seems more consistent with anthropology's mission to understand the whole range of human diversity than do either Heaven or Hell scenarios. The transhumanist aspiration to perfection—even becoming God-like—has little in common with Haraway's playful, ironic, and ambivalent sensibility in her launching of critical attention to the cyborg. She was suspicious of the limited capacity of reason to steer, much less optimize, what it produced. Skepticism about technological determinism also supports the idea that the consequences of the Singularity, if it occurs, and radical transformation more generally, will not follow a single track, either to Heaven or Hell, but will continue a trend for humanity collectively to muddle through its greatest trials, neither attaining perfection nor transgressing beyond redemption.

Readers may have noticed that we have not said much in this book about our own positions on the ethics of posthumanism, whether regarding animal rights or human enhancement, while noting the ethical disputes that others have raised. This reticence is primarily because we do not consider ourselves to have the training to productively engage in ethical philosophy. Many others have thought deeply about these issues (see the suggested readings in Chapter 3 for examples). Our interest here is more in how people make ethical claims, not to judge which claims are good, and how such claims influence the development of issues relevant to posthumanism. In that vein, we were intrigued to find that some of the richest engagement with transhumanist ideas has come from theologians. The first use of the term "transhumanism" was in 1957. Biologist Julian Huxley's book, entitled *Religion without Revelation*, discussed the possibility that the human species would transcend itself while remaining human by "realizing new possibilities of and for his human nature" (quoted in Peters 2015: 132).

Most transhumanists "see their movement as a replacement for traditional religion" (Peters 2015: 133). The enemy of the transhumanist is death: "Not only death, but also those who accept death, who advocate deathism. Who advocates deathism? Religious believers" (136). The central tenet of transhumanism can be summarized as the belief in overcoming human limitations through reason and technology. What more fundamental limitation is there than aging and death? Transhumanist optimism predicts average life expectancy exceeding a millennium. Ray Kurzweil (2005: 371) pronounces that we already have the means to "live long enough to live forever." Transhumanists offer two paths to this utopia. The first is radical life extension through genetics, nanotechnology, and robotics. This path remains biological. The second offers immortality through the uploading of consciousness.

This futurist dream raises major questions for theologians. If we can do such things, should we? Transhumanists tend to presume that religion will try to place roadblocks in their way because most religions are Luddite, dedicated to resisting change. Ted Peters (2015) points out that at least some theologians argue that there is no fundamental theological opposition to the project of uploading one's mind to a computer or other non-biological system. Immortality, though, would seem to raise serious questions—for example, about our need to worry about the afterlife. Is it sinful to avoid meeting your maker? Peters suggests that transhumanism leads down two separate paths: toward laissez-faire capitalism or toward increased cooperation and altruism. With the former, divides and inequities intensify. Potentially, the world might be divided among an elite of enhanced transhumans and those who are un-enhanced (either for inability to pay for it or ethical rejection) and find it impossible to compete in the labor market. This can be seen in the 1997 film *Gattaca*. The more recent film *Elysium* (2013) separated the elite from the poor by empty space, with the rich possessing remarkable medical technologies in luxurious orbital space stations, leaving the problems of earth behind, where the ordinary people struggled to survive and died from diseases that could easily be cured. Whatever the ethical conclusions might be, we can commend those who are beginning to seriously and openly engage with the questions that transhumanism poses for us in the decades and centuries to come.

Non-Western Transhumanism

It is important to note that the transhumanist movement continues to be dominantly Western in origin and inspiration. Non-Western versions of transhumanism have yet to receive much study, and the impression is that those proponents outside the West share an assumption that it simply represents a universalist aspiration for improvement. A Google and Google Scholar searches of "non-Western transhumanism" produced zero hits, generally a hard result to generate. While a Google Web search without the quotation marks, which finds the words without direct juxtaposition, did have more hits, few of these actually addressed non-Western approaches to transhumanist ideas, but were casual mentions of non-Western ideas, mostly with reference to the past. One useful source is a chapter by Heup Young Kim (2014), who addresses it from an East Asian theological context. He says that, as a theologian, his first reaction was to see transhumanism as one of the most dangerous ideas that the West has ever produced. On deeper study, he found it to resemble a "techno-secularization of the Judeo-Christian vision," including a technological version of the rapture, the transporting of believers to heaven at the Second Coming of Christ (99). From the perspective of non-monotheistic religions, he suggests that the

Enlightenment/transhumanist belief system is an "anthropocentrically reductionist worldview that has brought about ecological disaster by neglecting the holistic relationship of humans with the cosmos and the earth" (105). A neo-Confucianist position on transhumanism would build on a very different understanding of the commitments of the enlightened individual to nature and his or her fellow citizens, attempting to achieve benevolence rather than the will to power so evident in Western transhumanism. In a similar way, Ann Weinstone (2004) draws on Tantric Buddhism to engage with posthumanism, suggesting that the emphasis on human/non-human interaction has left the perspective with an impoverished ability to speak about human/human relationships. In *Tokyo Cyberpunk*, Steven Brown (2010: 159) says that because posthumanism is "profoundly transnational," there is no Japanese posthumanism as such, if that is defined as a discourse that is "specific to Japan alone."

So far, attention to technological enhancements outside the West is primarily still in the mode of Arturo Escobar's question (1994: 214): "What happens to non-Western perspectives as the new technologies extend their reach?" It seems very likely, however, that non-Western approaches to transhumanism will be less reactive and more important in the future, particularly given that a great deal of cutting-edge research in areas like robotics, stem cells, and aging is being done in East Asia.

As we stressed in Chapter 1, our approach to posthumanism contrasts greatly with transhumanism. It sees the past, not just the present and the future, as profoundly more than human. Recognizing the non-humans who have contributed to humanity's path toward becoming what it is now, in all its diversity around the world and in the cognisphere, perhaps helps us to think with a bit more balance and restraint about the epochal changes being proclaimed. Large changes do happen, but they happen through the complex patterns that have already been woven in the past and present by humans and their entangled others—microbial, animal, and technological. In emerging from current conditions, their character is profoundly influenced by the contexts in which they arise. Recognition of our posthuman past may help us to understand the strange futures that beckon toward us through the powerful technologies that promise or threaten to dramatically surpass our capabilities, and perhaps to do so without reliance on our supervision. Understanding transhuman possibilities is very important at this time in our history, but it is only one part of our more-than-human nature. Thinking about our posthuman past and present—recognizing how much diversity there is and has been in how we deal with our non-human companions and collaborators—may be the best foundation for plotting our course in what may become a Transanthropocene. What would a new geological era in which transhumans were the dominant shaping force look like, be like? What possibilities might it hold, what horrors? Cixin Liu (2001), a

Chinese science fiction author, offers one apocalyptic vision of humanity turning the earth into a vessel, and having to transform humanity to make this possible, in order to travel to Alpha Centauri after it is discovered that the sun will explode in 400 years. Less massive geo-engineering projects are being actively considered as responses to anthropogenic climate change. We have no answers to what the future might hold, but younger readers may need to address many such questions in the decades to come—or centuries, if the predictions of Kurzweil are correct.

Discussion and Activities

Consider several science fiction movies relevant to posthumanism (see some suggestions below). For each, ask if their perspective is transhumanist in the sense of assuming an Enlightenment view of human improvability, or if it is posthumanist in the sense used in this book, focusing on prosthetic extensions of human capacities in a more open or dystopic fashion.

Identify at least three ways in which your use of tools today has extended your capabilities. What difference do these capacity extensions make to your understanding of who and what you are?

Consider your use of social media. Does it create significant new ties with people for you, or do you use it primarily to interact with people you already know in the physical world?

Additional Readings and Films

Readings

Boellstorff, Tom. 2008. *Coming of Age in Second Life: An Anthropologist Explores the Virtually Human.* Princeton, NJ: Princeton University Press.

Bostrom, Nick. 2014. *Superintelligence: Paths, Dangers, Strategies.* Oxford: Oxford University Press.

Hakken, David. 2003. *The Knowledge Landscapes of Cyberspace.* London: Routledge.

Kurzweil, Ray. 2005. *The Singularity Is Near: When Humans Transcend Biology.* New York: Penguin.

Naam, Ramez. 2012. *Nexus.* Nottingham, UK: Angry Robot.

—. 2013. *Crux.* Nottingham, UK: Angry Robot.

—. 2015. *Apex.* Nottingham, UK: Angry Robot.

[A trilogy of novels about the consequences of ingested nanobots that provide direct connections between people's minds and make possible the emergence of transhumans.]

Scalzi, John. 2014. *Lock In.* New York: Tor Books.

[After a virus leaves one per cent of the world's population completely paralyzed but conscious, a research program produces robotic bodies with which they can control their minds at a distance.]

Films

A.I. Artificial Intelligence. 2001. Film directed by Steven Spielberg. Burbank, CA: Warner Bros.

Blade Runner. 1982. Film directed by Ridley Scott. California: The Ladd Company.

Edward Scissorhands. 1990. Film directed by Tim Burton. Los Angeles: Twentieth Century Fox.

Ex Machina. 2015. Film directed by Alex Garland. London: DNA Films.

Fixed: The Science/Fiction of Human Enhancement. Film directed by Regan Brashear. New York: New Day Films.

Flowers for Algernon. 2000. Film directed by Jeff Bleckner. Toronto: Alliance Atlantis Communications.

Gattaca. 1997. Film directed by Andrew Niccol. United States: Jersey Films.

Her. 2013. Film directed by Spike Jonze. Los Angeles: Annapurna Pictures.

Inception. 2010. Film directed by Christopher Nolan. Burbank, CA: Warner Bros.

Iron Man. 2008. Film directed by Jon Favreau. Los Angeles: Paramount Pictures.

Johnny Mnemonic. 1995. Film directed by Robert Longo. Culver City, CA: TriStar Pictures.

The Lawnmower Man. 1992. Film directed by Brett Leonard. United Kingdom: Allied Vision.

The Matrix. 1999. Film directed by The Wachowski Brothers. Burbank, CA: Warner Bros.

Metropolis. 1927. Film directed by Fritz Lang. Germany: Universum Film.

Minority Report. 2002 Film directed by Steven Spielberg. Los Angeles: Twentieth Century Fox.

WarGames. 1983. Film directed by John Badham. Los Angeles: United Artists.

Films Relevant to Prosthetic Extensions

Fat Man and Little Boy. 1989. Film directed by Roland Joffé. Los Angeles: Paramount Pictures.

The First Auto. 1927. Film directed by Roy Del Ruth. Burbank, CA: Warner Bros.

The Gift to Stalin. 2008. Film directed by Rustem Abdrashev. Kazakhstan: Aldongar Productions.

Modern Times. 1936. Film directed by Charles Chaplin. United States: Charles Chaplin Productions.

Person of Interest. 2011–2016. TV series directed by Jonathon Nolan et al. Beverly Hills: Kilter Films.

Quest for Fire. 1981. Film directed by Jean-Jacques Annaud. Canada: International Cinema Corporation.

Silk. 2007. Film directed by François Girard. Toronto: Rhombus Media.

The Social Network. 2010. Film directed by David Fincher. Culver City, CA: Columbia Pictures.

There Will Be Blood. 2007. Film directed by Paul Thomas Anderson. Los Angeles: Paramount Vantage.

The Truman Show. 1998. Film directed by Peter Weir. Los Angeles: Paramount Pictures.

Documentaries

America's Surveillance State. 2014. Documentary directed by Danny Schechter. United States: Globalvision.

Black Harvest. 1992. Documentary directed by Robin Anderson and Bob Connolly. Australia: Arundel Productions.

Bush Pilot: Reflections on a Canadian Myth. 1980. Documentary directed by Norma Bailey and Robert Lower. Montreal: National Film Board of Canada.

Cannibal Tours. 1988. Documentary directed by Dennis O'Rourke. Papua New Guinea: Institute of Papua New Guinea Studios.

Cyber Seniors. 2014. Documentary directed by Saffron Cassaday. Toronto and Los Angeles: The Best Part.

Drone. 2014. Documentary directed by Tonje Hessen Schei. Norway: Filmmer Film.

Manufacturing Consent: Noam Chomsky and the Media. 1992. Documentary directed by Mark Achbar and Peter Wintonick. Montreal: National Film Board of Canada.

McLuhan's Wake. 2002. Documentary directed by Kevin McMahon and David Sobelman. Montreal: National Film Board of Canada.

CONCLUSION

In many ways, the conclusion is possibly the most challenging part of a book for any author. All the key arguments and evidences have already been presented, debated, analyzed, expanded, and concluded in the main body of the manuscript. Instead of summarizing our key points one more time in this section, we use this space to explore further questions and possibilities about the human experience within possible future contexts of posthumanism and transhumanism. We wrote this book with the aim of informing readers, in easily understood language and with as little jargon as possible, that being human has always entailed being "more than human" through our embrace and incorporation of entities, organic and otherwise, within our bodies and living conditions. We consider that posthumanist perspectives have great insights to offer anyone interested in anthropology, or more generally in people, but that much of the writing in which the key ideas are presented is often daunting, even for specialists. Making these powerful and challenging ideas accessible to students and non-specialists motivated this writing project and helped us as well to clarify what it was that we found compelling in these ideas. One of our key aspirations is that readers will come to see themselves and their world in new ways. In this way, this book is itself a prosthesis, intended to shift our view and understanding through the application of the anthropological principle of holism in ways that challenge all the traditional boundaries of species and academic specialization.

Humans live and are renewed within an ever-changing mesh of entanglements that bind us to non-human elements by design and accident. How we undertake to intervene, modify, and live our coexistence with the greater universe of life forms, objects, and technologies will influence the paths of transformation of the human condition. We tend to think of ourselves, in a very ethnocentric and anthropocentric way, as the smartest life forms on earth. But are we? Judging by the fast deterioration of our living environments as a result of high carbon emissions and industrial pollutants rooted in excessive fossil fuel consumption and extraction, and the ineffective political and collective actions to solve such problems, we are as a species both the architect of our own misery and possible demise and the cause of the destruction of many other life forms. Our food system, guided by intense profit-seeking principles and enhanced by "scientific" knowledge, creates products that cause obesity and myriad other health problems. We are killing ourselves, literally. Perhaps there is no future for human existence on this planet if we keep going the way we have so far.

But we should not be so hasty as to write ourselves off. We have been incredibly successful in reproducing ourselves for a long time precisely because we are good at being more than human. There is an unfortunate rupture in our existence, in our opinion, when we believe in the supremacy of our technology and humanity—so much so that we see ourselves as the paramount reality outside of and above the rest of the life forms on this planet. We are not saying that technology is evil; on the contrary, technology is very important. But technology is a tool that requires human input, and how we use technology to achieve specific goals, usually with unintended consequences, is a complicated matter, tied to morality, ethics, and collective purpose. The use of exoskeletons to aid seniors in their mobility and the use of robotics to enhance their quality of life, in conjunction with ongoing medical advances to prolong life through surgery and medication, will create conditions that make it possible for more people to live productively to very old ages, past 100 or even 150. Will all future seniors benefit from these enhancements? Or will it create a new cleavage in human societies between the rich and the poor? Will such prolongation of life for the aged be achieved at the expense of services and employment opportunities for the young and the yet to be born? How could we feed and house all people in a future world where life expectancy carries no expiry dates?

Through this book, we hope to inspire readers to think more critically about our unhealthy and unproductive anthropocentric worldviews. We, *Homo sapiens*, have always lived and coevolved with other life forms. Meaningful and mutually beneficial coexistence requires respect and sensitivity for each other. Have we paid adequate attention to our interrelationships with the greater universe of life forms? How much do we really know about them? We have imposed our interpretation and understanding of other life forms in the manner of "what we think about the forest," and done too little to actively and respectfully learn about what forests are and how they do what they do. It is well past due that we focus on "how the forest thinks" (Kohn 2013). The long association of dogs and people is a wonderful illustration of our sentiment here that humans have not been good in learning about "how the forest thinks" or "how the dog thinks." We have, over time, manipulated the genetics of dogs to produce a variety of "breeds" that serve human desires and preferences. There are pit bulls to help with animal butchering; terriers to flush out underground animals; retrievers to bring back fallen birds; and working dogs, such as border collies, to herd sheep and cattle. We shape and change dogs into our image of what a working animal and companion animal should be

(Coulter 2016). We breed them increasingly for size, shape, and color. We create and modify these breeds of dogs without paying due attention to their physiology and capability, leading to the transmission of crippling genetic diseases (Haraway 2008). Today most dog owners, especially those in urban environments, know very little about the biology and capabilities of their companion animals. What a surprise it must be to the millions of dog owners to learn that the newest diagnostic tool for cancer is the dog. Quite by accident, a group of medical researchers in the United Kingdom confirmed in 2015 that some dogs have the ability to identify cancer patients even before clinical symptoms can be detected. It turns out the sebaceous (oil) secretions from the skin that adhere to the clothing of infected people can be detected by certain dogs. With early intervention, the chance of treating and surviving cancer can be greatly enhanced. If the offhand comment by the widow of a cancer patient that her dog acted differently with her husband during his illness was not taken seriously by a medical researcher on that fateful day, we might have had to wait many more years to learn about the diagnostic capability of dogs to identify cancer and possibly other health problems in the human body.

Given the well-established divisions in specialized research focuses and funding and disciplinary boundaries we have today, any attempt to fully investigate and explore the nature and consequences of the posthuman world would benefit from the practice of holism without boundaries. If we are to develop the potential for holism without boundaries, we need an inclusive multidisciplinary approach to foster meaningful joint research that complements each other's expertise among team members and a broader intellectual community of dialogue and respectful critique. The anthropological research method of participant observation is effective in generating new qualitative information about how people/animals/other life forms interact and the consequences of these interactions. Careful deployment of these methods can lead to new and unexpected insights, well before links like those between dogs and cancer diagnosis are made through other research methods. Once such links are considered as possibilities, validation of the hypothesis requires quantitative analysis, which can lead to new research questions that require inputs from experts in medicine, pharmacy, biochemistry, engineering, or ecology/environmental studies for a full and systematic investigation. This can lead to the development of techniques to optimize the effectiveness of the new possibilities for diagnosis or treatment. Of course, the order of events does not have to involve anthropologists first; it is at least as likely that medical experts or somebody else who listens to the observations of their patients or the public will raise new research questions first. The key point is that holism without boundaries is a methodological approach that can serve as a useful vantage point from which to learn about our (post)human condition, and the tangled web that being human in the Anthropocene, as well as in the past, has woven.

As we approached the end of this writing project, we learned that Professor Sidney Mintz (1922–2015) had just passed away. Sid was a dear friend whom we got to know on a personal level only much later in his life. His contribution to economic anthropology is tremendous, for which he received many awards, honors, and, naturally, much respect from a wide audience around the world. During one of our conversations some years ago, he mentioned that one main goal of anthropology is to ask, and hopefully to answer, the broader questions about the human experience and existence. In writing this book, we hope to make a small contribution toward this very noble but most ambitious goal of saying something meaningful about the human existence and condition. We are human, and we have always been more than human. May the force be with posthumanism and with the many collaborations that might help to search for escapes from the many dilemmas an anthropocentric worldview has created.

Some readers and students will still be wondering what we would tell them to do, given the complicated circumstances generated by the posthuman condition and the suggestion that we need to be less anthropocentric if we are to survive what we have brought into being. What should we eat, for example? How best could we use our more-than-human entanglements to save the planet? Within the context of this short book, we simply respond by quoting the rock band Ten Years After: "I'd love to change the world, but I don't know what to do, so I'll leave it up to you." We do hope that the ideas and examples presented in this book will give you some starting points and tools to find your own paths through the entangled problems of the Anthropocene.

Discussion and Activities

Imagine yourself with the capacity to change the world in some fundamental ways (perhaps you have just uploaded your consciousness and control the Internet). How would you try to improve the world? What non-humans would you need to recruit into your efforts? How might your interventions either have disastrous unintended consequences or be impeded by the character of the non-humans that you would need to use in your reform efforts?

GLOSSARY

Actants Entities, human or non-human, living or non-living, that make a difference in a controversy, action, or chain of events.

Actor–Network Theory (ANT) Diverse field of research that describes an actor, or actant, as anything that makes a difference in a network. Used particularly in studies of science.

Affordance A relationship between an object or an environment and an organism that affords the opportunity for that organism to perform an action.

Agency The ability to make things change, often assumed to involve intentionality or purpose behind the actions that an agent engages in.

Animism The worldview that sees animals, plants, and geological features like mountains possessing minds or souls, in contrast to the naturalist position that only humans have them.

Anthropocene The newly identified geological era in which human activity has become the dominant influence on climate and geological processes.

Anthropocentrism The tendency to put humans at the center of all considerations, seeing issues in terms of their implications for people.

Anthropomorphism The ascription of human qualities to non-humans.

Anthrozoology The study of interaction between humans and other animals.

Artificial intelligence (AI) The simulation, or potential achievement, of human-level intelligence processes by machines, especially computer systems.

Artificial life The field of study that examines systems related to life, its processes, and its evolution, through the use of simulations with computer models, robotics, and biochemistry.

Big data A term that describes a massive volume of structured and unstructured data that is so large that it is difficult to process using traditional database techniques.

Biosecurity Measures to reduce the risk of transmission of infectious diseases in crops and livestock, quarantined pests, invasive alien species, and biological warfare or terrorism.

Bovine spongiform encephalopathy (BSE) A fatal disease of cattle caused by infectious proteins (prions), which can be transmitted to humans through the consumption of brains and other risk materials.

Brain–computer interface A direct communication pathway between the brain and an external device, allowing the brain to control that device.

Coevolution The influence of associated species on each other in their evolution.

Cognitive schemas Patterns of thought or behavior that organize categories of information and the relationships among them.

Columbian exchange The exchange of diseases between New and Old Worlds resulting from European exploration and colonization. It caused devastation to populations without immunities.

Cultural relativism The position that ideas and behaviors are best seen in relation to the social context in which they take place. Interpreting them through outside ideas and worldviews introduces distortions and misunderstandings. Some anthropologists, but not all, take this as an ethical position.

Cybernetics The study and technology of communication and the control of machines and computers.

Cyberspace The space in which interactions between computers and their users occur, particularly through graphical representations.

Cyborg A human with implanted or attached cybernetic technology.

Cyborg anthropology The field of research into the incorporation of technology within human bodies and the addition of prosthetic extensions to those bodies, and the consideration of what this means for humans and our future.

Distributed cognition The ways in which cognitive resources are shared socially to extend individual cognitive resources or to accomplish something that an individual agent could not achieve alone.

Domestication The cultivating or taming of a population of animals or plants to produce or strengthen traits that are desirable to the cultivator or tamer.

Economies of scale The reduced cost per unit of a product that arises with increased output.

Ethnocentrism Belief in the superiority of one's own culture and a tendency to view other cultures from the perspective of one's own, often leading to misinterpretations.

Fecal transplants The transfer of feces from a donor to the bowel of a patient suffering from health problems related to the absence of microbes in their gut microbiome.

Food safety The conditions and practices that preserve the quality of food to prevent contamination and foodborne illnesses.

Holism The principle that parts of a society or culture are in intimate interconnection so that they cannot be understood without reference to the whole.

Human exceptionalism The worldview that, even when acknowledging that we are also animals, humans are much more than animals in what are thought to be the most important ways.

Humanism A worldview developed during the Enlightenment that emphasizes ideas of secularism, rationality, and the possibility of human progress and improvement.

Informationalization The addition of information to entities, products, and services that previously did not have such information enhancements.

Intentionality The quality of mental states such as thoughts, beliefs, and desires that involves their being directed toward some object or state of affairs.

Kuru A transmissible spongiform encephalopathies (TSEs) disease involving tremors, rapid deterioration, and death, first seen among the Fore of Papua New Guinea in the late 1920s.

Miasma theory The belief that disease was spread by bad air, widely held by physicians as well as the general population until late in nineteenth-century Europe.

Microbiome The microorganisms in a particular environment, including the human body.

Multiculturalism A policy that encourages the preservation of different cultures or cultural identities within a unified society.

Multinaturalism The worldview that assumes spiritual unity between humans and animals, accompanied by bodily differences, rather than the unity of nature and the plurality of cultures presumed by multiculturalism.

Multispecies ethnography Incorporating non-human animals into ethnographic field research.

Nanotechnology The science of producing particles and machines at the scale of less than 100 nanometers. A nanometer is one-billionth of a meter.

Naturalism The worldview associated with Western cultures that assumes that the same natural laws explain the physical characteristics of the world, so that they can be studied and understood by scientific methods.

Naturecultures A term that reflects the position of rejecting any dualistic division between nature and culture.

Pandemics Diseases prevalent throughout an entire country, continent, or the whole world.

Parasitism A relationship between two things in which one of them (the parasite) benefits from or lives off of the other.

Path dependency Tendency of a past practice or preference to continue even if better alternatives are available.

Polyvocality The use of multiple voices as a narrative mode within a text such as an ethnography.

Postcolonialism An academic framework that analyzes, explains, and responds to the cultural legacies of colonialism and imperialism.

Posthumanism As used in this book, an approach that rejects the anthropocentrism of humanism, and claims that as long as there have been humans, they have been posthuman in their reliance on non-humans.

Postmodernism Approaches that react against the asserted certainty of modernity, particularly in terms of scientific, or objective, efforts to explain reality or improve the world or human condition.

Post-structuralism Theoretical approaches that question the existence of essential features of any entity, and the coherence of the individual. People are instead seen as intersections of relationships, contexts, or actor-networks.

Probiotics Microorganism introduced into the body for their beneficial qualities.

Prosthetics A device, external or implanted, that substitutes for or supplements a part of the body.

Schizophrenia A mental illness involving a breakdown in the relationships between thought, emotion, and behavior, leading to faulty perception, inappropriate actions and feelings, and withdrawal from reality and personal relationships.

Singularity The transhumanist idea that, in the future, machine intelligence will surpass human intelligence, leading to a merger between human and artificial intelligence.

Speciesism The accusation, by analogy to racism and sexism, against those holding human exceptionalist ideas to reject the extension of rights to animals.

Technological determinism The idea that a technology directly causes certain kinds of change.

Transhumanism An intellectual movement that aims to transform the human condition by encouraging technologies to greatly enhance human intellectual, physical, and psychological capacities, producing transhumans (or H+) whose abilities are beyond those of current humans.

Zoonotic diseases Diseases that can be transmitted between animals and humans, also known as zoonoses.

REFERENCES

Anderson, Benedict. 2006. *Imagined Communities: Reflections on the Origin and Spread of Nationalism*. New York: Verso Books.

Appadurai, Arjun. 1996. *Modernity at Large: Cultural Dimensions of Globalization*. Minneapolis: University of Minnesota Press.

Beck, Ulrich. 1992. *Risk Society: Towards a New Modernity*. London: SAGE.

"Beetles and Bugs: Protecting Coffee Crops." 2015. *The Economist*, July 18: 68.

Bennett, Jane. 2009. *Vibrant Matter: A Political Ecology of Things*. Durham, NC: Duke University Press. http://dx.doi.org/10.1215/9780822391623.

Bestor, Theodore C. 2000. "How Sushi Went Global." *Foreign Policy* (121): 54–63. http://dx.doi.org/10.2307/1149619.

Bleed, Peter. 2006. "Living in the Human Niche." *Evolutionary Anthropology: Issues, News, and Reviews* 15 (1): 8–10. http://dx.doi.org/10.1002/evan.20084.

Bloch, Maurice E.F. 1998. *How We Think They Think: Anthropological Approaches to Cognition, Memory, and Literacy*. Boulder, CO: Westview Press.

Bloor, David. 1999. "Anti-Latour." *Studies in History and Philosophy of Science* 30 (1): 81–112. http://dx.doi.org/10.1016/S0039-3681(98)00038-7.

Boellstorff, Tom. 2008. *Coming of Age in Second Life: An Anthropologist Explores the Virtually Human*. Princeton, NJ: Princeton University Press.

Bostrom, Nick. 2005. "A History of Transhumanist Thought." *Journal of Evolution and Technology* 14 (1): 1–25.

Boyer, Dominic. 2012. "From Media Anthropology to the Anthropology of Mediation." In *The SAGE Handbook of Social Anthropology*, edited by R. Fardon, O. Harris, T.H. Marchand, C. Shore, V. Strang, R. Wilson, and M. Nuttall, 383–92. Thousand Oaks, CA: SAGE. http://dx.doi.org/10.4135/9781446201077.n66.

Braidotti, Rosi. 2013. *The Posthuman*. New York: John Wiley & Sons.

Brown, Steven T. 2010. *Tokyo Cyberpunk: Posthumanism in Japanese Visual Culture*. New York: Palgrave Macmillan. http://dx.doi.org/10.1057/9780230110069.

Bruinvels, G., R.J. Burden, A.J. McGregor, K.E. Ackerman, M. Dooley, T. Richards, and C. Pedlar. 2016. "Sport, Exercise and the Menstrual Cycle: Where Is the Research?" *British Journal of Sports Medicine*. http://dx.doi.org/10.1136/bjsports-2016-096279.

Busch, Lawrence. 2004. "Grades and Standards in the Social Construction of Safe Food." In *The Politics of Food*, edited by Marianne Elisabeth Lien and Brigitte Nerlich, 163–78. Oxford: Berg.

Calarco, Matthew. 2015. *Thinking through Animals: Identity, Difference, Indistinction.* Stanford, CA: Stanford University Press.

Callon, Michel. 1986. "Some Elements of a Sociology of Translation: Domestication of the Scallops and the Fishermen of St. Brieuc Bay." In *Power, Action, and Belief: A New Sociology of Knowledge*, edited by John Law, 196–223. London: Routledge.

Camporesi, Silvia. 2008. "Oscar Pistorius, Enhancement and Post-Humans." *Journal of Medical Ethics* 34 (9): 639. http://dx.doi.org/10.1136/jme.2008.026674.

Carter, Denise. 2005. "Living in Virtual Communities: An Ethnography of Human Relationships in Cyberspace." *Information Communication and Society* 8 (2): 148–67. http://dx.doi.org/10.1080/13691180500146235.

Childe, V. Gordon. 1936. *Man Makes Himself.* London: Watts and Co.

Clifford, James, and George E. Marcus, eds. 1986. *Writing Culture: The Poetics and Politics of Ethnography.* Berkeley: University of California Press.

Cole, Michael, and Jennifer Cole. 2006. "Rethinking the Goody Myth." In *Technology, Literacy, and the Evolution of Society: Implications of the Work of Jack Goody*, edited by David R. Olson and Michael Cole, 305–24. New York: Psychology Press.

Collier, Stephen J., and Andrew Lakoff. 2008. "The Problem of Securing Health." In *Biosecurity Interventions: Global Health and Security in Question*, edited by Andrew Lakoff and Stephen J. Collier, 7–32. New York: Columbia University Press. http://dx.doi.org/10.7312/lako14606-001.

Collins, Harry M., and Trevor Pinch. 1998. *The Golem: What You Should Know about Science.* 2nd ed. New York: Cambridge University Press.

Core Jr. 2012. "Case Study: Ento, the Art of Eating Insects." *Core77*, February 27. http://www.core77.com/posts/21841/case-study-ento-the-art-of-eating-insects-21841.

Coulter, Kendra. 2016. *Animals, Work, and the Promise of Interspecies Solidarity.* New York: Palgrave MacMillan. http://dx.doi.org/10.1057/9781137558800.

Crisolago, Mike. 2015. "This Way Up." *Zoomer* 31 (9): 15.

Cryan, J.F., and S.M. O'Mahony. 2011. "The Microbiome-Gut-Brain Axis: From Bowel to Behavior." *Neurogastroenterology and Motility* 23 (3): 187–92. http://dx.doi.org/10.1111/j.1365-2982.2010.01664.x.

da Silva, Rodrigo Costa, and Helio Langoni. 2009. "*Toxoplasma gondii*: Host–Parasite Interaction and Behavior Manipulation." *Parasitology Research* 105 (4): 893–98. http://dx.doi.org/10.1007/s00436-009-1526-6.

Davies, Tony. 1997. *Humanism: The New Critical Idiom.* London: Routledge.

de Castro, Eduardo Viveiros. 1998. "Cosmological Deixis and Amerindian Perspectivism." *Journal of the Royal Anthropological Institute* 4 (3): 469–88. http://dx.doi.org/10.2307/3034157.

De Waal, Frans B.M. 2008. "Putting the Altruism Back into Altruism: The Evolution of Empathy." *Annual Review of Psychology* 59 (1): 279–300.

DeMello, Margo. 2012. *Animals and Society: An Introduction to Human-Animal Studies.* New York: Columbia University Press.

Descola, Philippe. 2013. *Beyond Nature and Culture.* Chicago: University of Chicago Press.

DiNovelli-Lang, Danielle. 2013. "The Return of the Animal: Posthumanism, Indigeneity, and Anthropology." *Environmental Sciences* 4 (1): 137–56. http://dx.doi.org/10.3167/ares.2013.040109.

Downey, Gary Lee, Joseph Dumit, and Sarah Williams. 1995. "Cyborg Anthropology." *Cultural Anthropology* 10 (2): 264–69. http://dx.doi.org/10.1525/can.1995.10.2.02a00060.

Dumit, Joseph, and Robbie Davis-Floyd. 1998. "Introduction: Cyborg Babies, Children of the Millennium." In *Cyborg Babies: From Techno-Sex to Techno-Tots*, edited by Robbie Davis-Floyd and Joseph Dumit, 1–18. New York: Routledge.

Dunn, Fred J. 2004. *Report of the Auditor General on the Alberta Government's BSE-Related Assistance Programs*. Edmonton: Office of the Auditor General. https://www.oag.ab.ca/node/58.

Dunn, Rob R. 2011. *The Wild Life of Our Bodies: Predators, Parasites, and Partners that Shape Who We Are Today*. New York: Harper.

Dunworth, Treas. 2009. "Biosecurity in New Zealand." In *Biosecurity: Origins, Transformations and Practices*, edited by Brian Rappert and Chandre Gould, 156–70. Houndmills, UK: Palgrave Macmillan. http://dx.doi.org/10.1057/9780230245730_9.

Edwards, Jeanette, Sarah Franklin, Eric Hirsch, Frances Price, and Marilyn Strathern, eds. 1999. *Technologies of Procreation: Kinship in the Age of Assisted Conception*. London: Routledge.

Escobar, Arturo. 1994. "Welcome to Cyberia: Notes on the Anthropology of Cyberculture." *Current Anthropology* 35 (3): 211–31. http://dx.doi.org/10.1086/204266.

Farmer, Paul. 2001. *Infections and Inequalities: The Modern Plagues*. Berkeley: University of California Press.

Fedigan, Linda Marie. 2001. "The Paradox of Feminist Primatology: The Goddess's Discipline." In *Feminism in Twentieth-Century Science, Technology and Medicine*, edited by Angela N. H. Crenger, Elizabeth Lunbeck, and Londa Schiebirger, 46–72. Chicago: University of Chicago Press.

Fidler, David P. 2010. "Towards a Global *Ius Pestilentiae*: The Functions of Law in Global Biosecurity." In *Global Biosecurity: Threats and Responses*, edited by Peter Katona, John P. Sullivan, and Michael D. Intriligator, 286–302. London: Routledge.

Flegr, Jaroslav. 2007. "Effects of Toxoplasma on Human Behavior." *Schizophrenia Bulletin* 33 (3): 757–60. http://dx.doi.org/10.1093/schbul/sbl074.

Foner, Nancy. 2000. *From Ellis Island to JFK: New York's Two Great Waves of Immigration*. New Haven, CT: Yale University Press.

Franklin, Sarah. 2013. *Biological Relatives-IVF, Stem Cells and the Future of Kinship*. Durham, NC: Duke University Press. http://dx.doi.org/10.1215/9780822378259.

Freidberg, Susanne. 2004. *French Beans and Food Scares: Culture and Commerce in an Anxious Age*. New York: Oxford University Press.

Gandy, Matthew. 2005. "Cyborg Urbanization: Complexity and Monstrosity in the Contemporary City." *International Journal of Urban and Regional Research* 29 (1): 26–49. http://dx.doi.org/10.1111/j.1468-2427.2005.00568.x.

Garreau, Joel. 2006. *Radical Evolution: The Promise and Peril of Enhancing Our Minds, Our Bodies—and What It Means to Be Human*. New York: Broadway.

Geertz, Clifford. 1963. *Agricultural Involution: The Process of Ecological Change in Indonesia.* Berkeley: University of California Press.

—. 1973. *The Interpretation of Cultures: Selected Essays.* New York: Basic Books.

"Genome Editing: The Age of the Red Pen." 2015. *The Economist*, August 22: 19–22.

Ghitter, Geoff, and Alan Smart. 2008. "Mad Cows, Regional Governance and Urban Sprawl: Path Dependence and Unintended Consequences in the Calgary Region." *Urban Affairs Review* 44 (5): 617–44. http://dx.doi.org/10.1177/1078087408325257.

Gibson, Hannah. 2015. "Exploring Contemporary Anthropological Theory: Can We Use Posthumanism to Reconceptualise the Disabled Body?" *Sites: A Journal of Social Anthropology and Cultural Studies* 12 (2): 3–21. https://sites.otago.ac.nz/Sites/article/view/297/314http://dx.doi.org/10.11157/sites-vol12iss2id297.

Gibson, James Jerome. 1979. *The Ecological Approach to Visual Perception.* Boston: Houghton Mifflin.

Ginsburg, Faye, and Rayna Rapp. 2013. "Entangled Ethnography: Imagining a Future for Young Adults with Learning Disabilities." *Social Science & Medicine* 99: 187–93. http://dx.doi.org/10.1016/j.socscimed.2013.11.015.

Goody, Jack. 2010. *Renaissances: The One or the Many?* Cambridge: Cambridge University Press.

Goody, Jack, and Ian Watt. 1963. "The Consequences of Literacy." *Comparative Studies in Society and History* 5 (3): 304–45. http://dx.doi.org/10.1017/S0010417500001730.

Grimm, David. 2014. *Citizen Canine: Our Evolving Relationship with Cats and Dogs.* New York: Public Affairs.

Hakken, David. 1993. "Computing and Social Change: New Technology and Workplace Transformation, 1980–1990." *Annual Review of Anthropology* 22 (1): 107–32. http://dx.doi.org/10.1146/annurev.an.22.100193.000543.

—. 2003. *The Knowledge Landscapes of Cyberspace.* London: Routledge.

—. 2004. "The Cyberspace Anthropology: A Foreword." *Antropologi Indonesia* 73: iv–xvi.

Hall, Stuart. 1977. "Culture, the Media and the 'Ideological Effect,'" In *Mass Communication and Society*, edited by J. Curran, M. Gurevitch, and J. Woolacott, 315–48. London: Edward Arnold.

Halverson, John. 1992. "Goody and the Implosion of the Literacy Thesis." *Man* 27 (2): 301–17. http://dx.doi.org/10.2307/2804055.

Hansen, Paul. 2013. "Urban Japan's 'Fuzzy' New Families: Affect and Embodiment in Dog–Human Relationships." *Asian Anthropology* 12 (2): 83–103. http://dx.doi.org/10.1080/1683478X.2013.852718.

Haraway, Donna. 1985. "Manifesto for Cyborgs: Science, Technology, and Socialist Feminism in the 1980s." *Socalist Review* 80: 65–108.

—. 1989. *Primate Visions.* New York: Routledge.

—. 1991. *Simians, Cyborgs and Women: The Reinvention of Nature.* New York: Routledge.

—. 2003. *The Companion Species Manifesto: Dogs, People, and Significant Otherness.* Chicago: Prickly Paradigm Press.

—. 2008. *When Species Meet.* Minneapolis: University of Minnesota Press.

Hayles, N. Katherine. 1999. *How We Became Posthuman: Virtual Bodies in Cybernetics, Literature, and Informatics*. Chicago: University of Chicago Press. http://dx.doi.org/10.7208/chicago/9780226321394.001.0001.

—. 2006. "Unfinished Work from Cyborg to Cognisphere." *Theory, Culture & Society* 23 (7–8): 159–66.

Helmreich, Stefan. 1998. *Silicon Second Nature: Culturing Artificial Life in a Digital World*. Berkeley: University of California Press.

Hinchliffe, Steve, and Kath Woodward. 2004. *The Natural and the Social: Uncertainty, Risk, Change*. 2nd ed. New York: Routledge.

Hodder, Ian. 2012. *Entangled: An Archaeology of the Relationships between Humans and Things*. London: John Wiley & Sons. http://dx.doi.org/10.1002/9781118241912.

—. 2014. "The Entanglements of Humans and Things: A Long-Term View." *New Literary History* 45 (1): 19–36. http://dx.doi.org/10.1353/nlh.2014.0005.

Honda, Kenya, and Dan R. Littman. 2012. "The Microbiome in Infectious Disease and Inflammation." *Annual Review of Immunology* 30 (1): 759–95. http://dx.doi.org/10.1146/annurev-immunol-020711-074937.

Hughes, David P., Sandra B. Andersen, Nigel L. Hywel-Jones, Winanda Himaman, Johan Billen, and Jacobus J Boomsma. 2011. "Behavioral Mechanisms and Morphological Symptoms of Zombie Ants Dying from Fungal Infection." *BMC Ecology* 11 (13). http://dx.doi.org/10.1186/1472-6785-11-13.

Hurn, Samantha. 2012. *Humans and Other Animals: Cross-Cultural Perspectives on Human-Animal Interactions*. London: Pluto Press.

Ingold, Tim. 1994. "Humanity and Animality." In *Companion Encyclopedia of Anthropology: Humanity, Culture and Social Life*, edited by Tim Ingold, 14–32. New York: Routledge.

—. 2000. *The Perception of the Environment: Essays on Livelihood, Dwelling and Skill*. London: Routledge.

—. 2008. "Bindings against Boundaries: Entanglements of Life in an Open World." *Environment & Planning A* 40 (8): 1796–810. http://dx.doi.org/10.1068/a40156.

—. 2011. *Being Alive: Essays on Movement, Knowledge and Description*. New York: Routledge.

—. 2013. "Prospect." In *Biosocial Becomings: Integrating Social and Biological Anthropology*, edited by Tim Ingold and Gisli Palsson, 1–21. New York: Cambridge University Press. http://dx.doi.org/10.1017/CBO9781139198394.002.

Jacobus, Mary, Evelyn Fox Keller, and Sally Shuttleworth, eds. 1990. *Body/Politics: Women and the Discourses of Science*. New York: Routledge.

Jamison, Douglas W. 2014. "Technologies that Will Drive Innovation and Investing in 2015." *Harris & Harris Group Blog*, October 14. http://www.hhvc.com/blog/technologies-will-drive-innovation-investing-2015/#sthash.DI2ojHYT.dpuf.

Joy, Melanie. 2009. *Why We Love Dogs, Eat Pigs, and Wear Cows: An Introduction to Carnism*. San Francisco: Conari Press.

Kim, Heup Young. 2014. "Cyborg, Sage, and Saint: Transhumanism as Seen from an East Asian Theological Setting." In *Religion and Transhumanism: The Unknown Future of Human Enhancement*, edited by Calvin Mercer and Tracy J. Trothen, 97–114. Santa Barbara, CA: Praeger.

Kirksey, Eben, ed. 2014. *The Multispecies Salon*. Durham, NC: Duke University Press. http://dx.doi.org/10.1215/9780822376989.

Kirksey, S. Eben, and Stefan Helmreich. 2010. "The Emergence of Multispecies Ethnography." *Cultural Anthropology* 25 (4): 545–76. http://dx.doi.org/10.1111/j.1548-1360.2010.01069.x.

Knauft, Bruce M. 1996. *Genealogies for the Present in Cultural Anthropology*. New York: Routledge.

Kohn, Eduardo. 2007. "How Dogs Dream: Amazonian Natures and the Politics of Transspecies Engagement." *American Ethnologist* 34 (1): 3–24. http://dx.doi.org/10.1525/ae.2007.34.1.3.

———. 2013. *How Forests Think: Toward an Anthropology Beyond the Human*. Berkeley: University of California Press.

Kuhn, Thomas S.T. 2012. *The Structure of Scientific Revolutions*. Chicago: University of Chicago Press. http://dx.doi.org/10.7208/chicago/9780226458144.001.0001.

Kurzweil, Ray. 2005. *The Singularity Is Near: When Humans Transcend Biology*. New York: Penguin.

Lakoff, Andrew. 2008. "From Population to Vital System." In *Biosecurity Interventions: Global Health and Security in Question*, edited by Andrew Lakoff and Stephen J. Collier, 33–60. New York: Columbia University Press. http://dx.doi.org/10.7312/lako14606-002.

Lanier, Jaron. 2010. *You Are Not a Gadget*. New York: Knopf.

Larkin, Brian. 2008. *Signal and Noise: Media, Infrastructure, and Urban Culture in Nigeria*. Durham, NC: Duke University Press. http://dx.doi.org/10.1215/9780822389316.

Lash, Scott. 1990. "Postmodernism as Humanism? Urban Space as Social Theory." In *Theories of Modernity and Postmodernity*, edited by Bryan Turner, 44–51. Thousand Oaks, CA: SAGE.

Latour, Bruno. 1987. *Science in Action: How to Follow Scientists and Engineers through Society*. Cambridge, MA: Harvard University Press.

———. 1993. *We Have Never Been Modern*. Cambridge, MA: Harvard University Press.

———. 1999. "For David Bloor…and Beyond: A Reply to David Bloor's 'Anti-Latour.'" *Studies in History and Philosophy of Science* 30 (1): 113–29.

———. 2005. *Reassembling the Social*. London: Oxford University Press.

Le Roy, Danny G., and K.K. Klein. 2005. "Mad Cow Chaos in Canada: Was It Just Bad Luck or Did Government Policies Play a Role?" *Canadian Public Policy* 31 (4): 381–99. http://dx.doi.org/10.2307/3552357.

Leach, Edmund. 1976. *Culture and Communication*. Cambridge: Cambridge University Press. http://dx.doi.org/10.1017/CBO9780511607684.

Leitner, Gerhard. 2015. *The Future Home Is Wise, Not Smart: A Human-Centric Perspective on Next Generation Domestic Technologies*. Cham, Switzerland: Springer. http://dx.doi.org/10.1007/978-3-319-23093-1.

Lien, Marianne Elisabeth. 2004. "The Politics of Food: An Introduction." In *The Politics of Food*, edited by Marianne Elisabeth Lien and Brigitte Nerlich, 1–18. New York: Berg.

Lindenbaum, Shirley. 2001. "Kuru, Prions, and Human Affairs: Thinking about Epidemics." *Annual Review of Anthropology* 30 (1): 363–85. http://dx.doi.org/10.1146/annurev.anthro.30.1.363.

———. 2008. "Understanding Kuru: The Contribution of Anthropology and Medicine." *Philosophical Transactions of the Royal Society of London. Series B, Biological Sciences* 363 (1510): 3715–20. http://dx.doi.org/10.1098/rstb.2008.0072.

Liu, Cixin. 2001. *The Wandering Earth*. Beijing: Guomi Digital.

Lorimer, Jamie. 2015. *Wildlife in the Anthropocene*. Minneapolis: University of Minnesota Press. http://dx.doi.org/10.5749/minnesota/9780816681075.001.0001.

Macbeth, Helen, and Sue Lawry. 1997. "Introduction." In *Food Preferences and Taste: Continuity and Change*, edited by Helen Macbeth, 1–14. New York: Berghahn Books.

Marcus, George. 2010. "Holism and the Expectations of Critique in Post-1980s Anthropology." In *Experiments in Holism: Theory and Practice in Contemporary Anthropology*, edited by Ton Otto and Nils Bubandt, 28–46. New York: Wiley-Blackwell. http://dx.doi.org/10.1002/9781444324426.ch3.

Marcus, George E., and Michael M. Fischer. 1986. *Anthropology as Cultural Critique: An Experimental Moment in the Human Sciences*. Chicago: University of Chicago Press.

Maturana, Humberto R., and Francisco J. Varela. 1972. *Autopoiesis and Cognition: The Realization of the Living*. Dordrecht, Netherlands: D. Reidel.

Michaels, Eric. 1987. "Hollywood Iconography: A Warlpiri Reading." In *Television and Its Audience: International Research Perspectives*, edited by P. Drummond and R. Patterson, 81–95. London: BFI.

Miller, Daniel, and Heather A. Horst. 2012. "The Digital and the Human: Prospectus for Digital Anthropology." In *Digital Anthropology*, edited by Heather A. Horst and Daniel Miller, 3–35. London: Berg.

Miller, William Ian. 1997. *The Anatomy of Disgust*. Cambridge, MA: Harvard University Press.

Minaei, Negin. 2014. "Do Modes of Transportation and GPS Affect Cognitive Maps of Londoners?" *Transportation Research Part A, Policy and Practice* 70: 162–80. http://dx.doi.org/10.1016/j.tra.2014.10.008.

Naam, Ramez. 2005. *More Than Human: Embracing the Promise of Biological Enhancement*. New York: Broadway Books.

Nading, Alex M. 2014. *Mosquito Trails: Ecology, Health, and the Politics of Entanglement*. Berkeley: University of California Press. http://dx.doi.org/10.1525/california/9780520282612.001.0001.

Nauert Jr., Charles G. 1995. *Humanism and the Culture of Renaissance Europe*. Vol. 6, *New Approaches to European History*. Cambridge: Cambridge University Press.

Nunn, Nathan, and Nancy Qian. 2010. "The Columbian Exchange: A History of Disease, Food, and Ideas." *Journal of Economic Perspectives* 24 (2): 163–88. http://dx.doi.org/10.1257/jep.24.2.163.

O'Connor, Terence P. 1997. "Working at Relationships: Another Look at Animal Domestication." *Antiquity* 71 (271): 149–56. http://dx.doi.org/10.1017/S0003598X00084635.

Otto, Ton, and Nils Bubandt, eds. 2010. *Experiments in Holism: Anthropology and the Predicaments of Holism*. New York: Wiley-Blackwell. http://dx.doi.org/10.1002/9781444324426.

Peters, Ted. 2015. "Theologians Testing Transhumanism." *Theology and Science* 13 (2): 130–49. http://dx.doi.org/10.1080/14746700.2015.1023524.

Pickering, Andrew. 1995. *The Mangle of Practice: Time, Agency, and Science*. Chicago: University of Chicago Press. http://dx.doi.org/10.7208/chicago/9780226668253.001.0001.

—. 2010. *The Cybernetic Brain: Sketches of Another Future*. Chicago: University of Chicago Press.

—. 2013. "Living in the Material World." In *Materiality and Space: Organizations, Artefacts and Practices*, edited by Nathalie Mitev and Francois-Xavier de Vaujany, 25–40. London: Palgrave Macmillan. http://dx.doi.org/10.1057/9781137304094_2.

Pink, Sarah, Kerstin Leder Mackley, Val Mitchell, Garrath T. Wilson, and Tracy Bhamra. 2016. "Refiguring Digital Interventions for Energy Demand Reduction: Designing for Life in the Digital-Material Home." In *Digital Materialities: Design and Anthropology*, edited by Sarah Pink, Elisenda Ardevol, and Debora Lanzeni, 79–98. London: Bloomsbury.

Porter, Dorothy. 1999. *Health, Civilization, and the State: A History of Public Health from Ancient to Modern Times*. London: Routledge.

Prusiner, Stanley. 1997. "Prion Diseases and the BSE Crisis." *Science* 278 (5336): 245–51. http://dx.doi.org/10.1126/science.278.5336.245.

Putnam, Tim. 2006. "Postmodern Home Life." In *At Home: An Anthropology of Domestic Space*, edited by Irene Cieraad, 145–52. New York: Syracuse University Press.

Quammen, David. 2012. *Spillover: Animal Infections and the Next Human Pandemic*. New York: W.W. Norton & Company.

Rampton, Sheldon, and John Stauber. 2004. *Mad Cow U.S.A*. Monroe, ME: Common Courage Press.

Rapp, Rayna. 2004. *Testing Women, Testing the Fetus: The Social Impact of Amniocentesis in America*. New York: Routledge.

—. 2015. "Big Data, Small Kids: Medico-Scientific, Familial and Advocacy Visions of Human Brains." *BioSocieties* 11 (3): 296–316. http://dx.doi.org/10.1057/biosoc.2015.33.

Roberts, Elizabeth F.S. 2012. *God's Laboratory: Assisted Reproduction in the Andes*. Berkeley: University of California Press.

Robinson, Charlotte, Clara Mancini, Janet van der Linden, Claire Guest, and Rob Harris. 2014. "Canine-Centered Interface Design: Supporting the Work of Diabetes Alert Dogs." In *ACM CHI'14 Proceedings of the SIGCHI Conference on Human Factors in Computing Systems, 26 April, Toronto, Canada*, 3757–766. New York: ACM. http://dx.doi.org/10.1145/2556288.2557396.

Roseberry, William. 1982. "Balinese Cockfights and the Seduction of Anthropology." *Social Research* 49 (4): 1013–28.

"Saccharin Solution?" 2014. *The Economist*, September 20: 74–75.

Sahlins, Marshall. 1978. *Culture and Practical Reason*. Chicago: University of Chicago Press.

Sample, Ian. 2015. "Summit Rules Out Ban on Gene Editing Embryos Destined to Become People." *The Guardian*, December 3. https://www.theguardian.com/science/2015/dec/03/gene-editing-summit-rules-out-ban-on-embryos-destined-to-become-people-dna-human.

Skogstad, Grace. 2006. "Regulating Food Safety Risks in the European Union: A Comparative Perspective." In *What's the Beef? The Contested Governance of European Food Safety*, edited by Christopher Ansell and David Vogel, 213–36. Cambridge, MA: MIT Press.

Smart, Alan. 2011. "The Humanism of Postmodernist Anthropology and the Post-Structuralist Challenges of Posthumanism." *Anthropologica* 53 (2): 332–34.

———. 2014. "Critical Perspectives on Multispecies Ethnography." *Critique of Anthropology* 34 (1): 3–7.

Smart, Alan, and Josephine Smart. 2008. "Time-Space Punctuation: Hong Kong's Border Regime and Limits on Mobility." *Pacific Affairs* 81 (2): 175–93. http://dx.doi.org/10.5509/2008812175.

———. 2012a. "Quarantine, Biosecurity and Life across the Border." In *A Companion to Border Studies*, edited by Thomas M. Wilson and Hastings Donnan, 354–70. New York: Wiley-Blackwell.

———. 2012b. "Governing Beef: Program Implementation, Unintended Consequences, and BSE Control in Alberta." In *Governing Cultures: Anthropological Perspectives on Political Labor, Power, and Government*, edited by Kendra Coulter and William R. Schumann, 69–92. New York: Palgrave Macmillan.

———. 2016. "Tangled Up in Food: The Moral Economy of Food Politics in the Transpacific Region." In *Transpacific Americas: Encounters and Engagements Between the Americas and the South Pacific*, edited by Eveline Dürr and Philipp Schorch, 150–66. New York: Routledge.

Smart, Josephine. 2008. "Will the Alberta Livestock and Meat Strategy Serve the Best Interests of Cattle Producers?" *Two Hills and County Chronicle*, August 12: 3.

Solomon, Olga. 2012. "Doing, Being and Becoming: The Sociality of Children with Autism in Activities with Therapy Dogs and Other People." *Cambridge Anthropology* 30 (1): 109–26.

Spriggs, John, and Grant Isaac. 2001. *Food Safety and International Competitiveness: The Case of Beef*. Wallingford, UK: CABI Publishing. http://dx.doi.org/10.1079/9780851995182.0000.

Stephenson, Neal. 2015. *Seveneves: A Novel*. New York: William Morrow.

"Stock Answers." 2016. *The Economist*, June 11: TQ15.

Strengers, Yolande. 2016. "Envisioning the Smart Home: Reimagining a Smart Energy Future." In *Digital Materialities: Design and Anthropology*, edited by Sarah Pink, Elisenda Ardevol, and Debora Lanzeni, 61–76. London: Bloomsbury.

Torfing, Jacob. 1999. "Towards a Schumpeterian Workfare Postnational Regime: Path-Shaping and Path-Dependency in Danish Welfare State Reform." *Economy and Society* 28 (3): 369–402. http://dx.doi.org/10.1080/03085149900000010.

Tsing, Anna. 2010. "Worlding the Matsutake Diaspora: Or, Can Actor-Network Theory Experiment with Holism?" In *Experiments in Holism: Anthropology and the Predicaments of Holism*, edited by Ton Otto and Nils Bubandt, 47–66. New York: Wiley-Blackwell. http://dx.doi.org/10.1002/9781444324426.ch4.

—. 2012. "Unruly Edges: Mushrooms as Companion Species." *Environmental Humanities* 1 (1): 141–54. http://dx.doi.org/10.1215/22011919-3610012.

—. 2015. *The Mushroom at the End of the World: On the Possibility of Life in Capitalist Ruins*. Princeton, NJ: Princeton University Press.

Tufekci, Zeynep. 2008. "Grooming, Gossip, Facebook and MySpace: What Can We Learn about These Sites from Those Who Won't Assimilate?" *Information Communication and Society* 11 (4): 544–64. http://dx.doi.org/10.1080/13691180801999050.

Turnbaugh, Peter J., Ruth E. Ley, Micah Hamady, Claire Fraser-Liggett, Rob Knight, and Jeffrey I. Gordon. 2007. "The Human Microbiome Project." *Nature* 449 (7164): 804–10. http://dx.doi.org/10.1038/nature06244.

Udell, Monique A.R., Nicole R. Dorey, and Clive D.L. Wynne. 2010. "What Did Domestication Do to Dogs? A New Account of Dogs' Sensitivity to Human Actions." *Biological Reviews of the Cambridge Philosophical Society* 85 (2): 327–45. http://dx.doi.org/10.1111/j.1469-185X.2009.00104.x.

Venturini, Tommaso. 2010. "Diving in Magma: How to Explore Controversies with Actor-Network Theory." *Public Understanding of Science* 19 (3): 258–73. http://dx.doi.org/10.1177/0963662509102694.

Watts, Sheldon. 1997. *Epidemics and History: Disease, Power and Imperialism*. New York: Yale University Press.

Weinstone, Ann. 2004. *Avatar Bodies: A Tantra for Posthumanism*. Minneapolis: University of Minnesota Press.

"Where the Smart Is: Connected Homes Will Take Longer to Materialise than Expected." 2016. *The Economist*, June 11: 65–66.

Wilson, Charlie, Tom Hargreaves, and Richard Hauxwell-Baldwin. 2015. "Smart Homes and Their Users: A Systematic Analysis and Key Challenge." *Personal and Ubiquitous Computing* 19 (2): 463–76. http://dx.doi.org/10.1007/s00779-014-0813-0.

Wolfe, Cary. 2010. *What Is Posthumanism?* Minneapolis: University of Minnesota Press.

—. 2012. *Before the Law: Humans and Other Animals in a Biopolitical Frame*. Chicago: University of Chicago Press.

Wrangham, Richard. 2009. *Catching Fire: How Cooking Made Us Human*. New York: Basic Books.

Zimmer, Carl. 2012. *A Planet of Viruses*. Chicago: University of Chicago Press.

INDEX

A

actants, in ANT and agency, 23, 24–27, 34, 49, 71
Actor–Network Theory (ANT)
   actants and agency, 23, 24–27, 34, 49, 71
   and BSE crisis, 38
   and controversies, 34–35
   criticism of, 23, 24
   description, 22–23
   hybrids in, 24
affordance, 72
agency
   in ANT, 24–26
   definition, 23
   and intentionality, 24, 26
   in material realm, 26, 30–31, 41
   and non-humans, 24–25, 26–27, 29–30
   and science, 29–31, 49
agriculture, 57–60, 84
Alberta, BSE crisis, 11, 39–40
Amazonian peoples, multinaturalism, 47
Amerindians, multinaturalism, 47
Anderson, Benedict, 73–74
Animal–human relations, 5, 55–57
animality, 55
animals
   in Amerindian thought, 47
   classification, 60
   in cosmologies, 47–48
   and culture, 12, 55
   as cyborgs, 83–84
   domestication, 57–58
   as food, 59–62
   as furry children, 58–59
   humanity and dualism in, 4–5, 43–46, 47–48, 49
   humans as, 44, 45–46
   Western worldviews, 12, 45–46
   See also non-humans
animism, 47–48, 49
ANT. *See* Actor–Network Theory (ANT)

Anthropocene, 3, 34
anthropocentrism
   and Anthropocene, 3
   rejection of and post-anthropocentric path, 4–6, 12, 29, 48, 62, 96, 98
anthropology
   culture in, 52–53, 54–55
   definitions of, 1, 6
   divisions in, 53
   entanglements and holism, 7–8, 9–10, 23, 95–96
   evolution of, 6–7, 51–52
   fields in, 65
   methods and perspectives, 43, 50–52, 53
   non-humans in, 50, 52–53, 54–55, 56
   and posthumanism, 6, 9, 52
   and postmodernism, 50, 51–53, 54–55
   scope and mission, 6–7
   and small-scale societies, 9, 23, 51
   and technology, 1–2
   worldviews in, 52–53
anthropomorphism, 59
anthrozoology, 5, 55–57
antibiotic resistance, 40
apes, 69
Appadurai, Arjun, 73
artificial intelligence (AI), 85

B

babies, 19, 80
Bali (Indonesia), 55
behavior, in microbiota, 21
Bennett, Henry, 36
bioconservatism, 87
biological weapons and bioterrorism, 32
biosecurity, 17, 32–33
"biosocial becomings", 49
"black box" idea, 35
Bleed, Peter, 57
Bloch, Maurice, 70
body, views on, 47, 48

body–technology mergers. *See* transhumanism
borders, 10–11, 17, 31–33
Bostrom, Nick, 86–87
boundaries
    and holism, 7–9, 11
    holism without, 9, 13, 65, 97–98
brain (human), 1–2, 21, 79–80, 85
brain–computer interface, 67
Brown, Steven, 91
BSE and crisis (bovine spongiform encephalopathy)
    background research, 35–37
    in Canada, 11, 39–40
    cause of, 37
    consequences, 11, 38–41
    and controversy, 34–35
    description, 35–36
    development and spread, 37–38
    and ear-tags, 84
    food safety and regulation, 38–39
    in food trade, 40–41
    and prions, 37, 38
Bush, George W., 87

C
Callon, Michel, 25, 48
cancer and dogs, 97
cannibalism, 36–37, 60
Carter, Denise, 75
chickens in Bali, 55
Childe, V. Gordon, 57
China, entanglements in SARS, 8
climate change, as hybrid, 24
Clostridium difficile, and fecal transplants, 22
cockfighting in Bali, 54–55
coevolution, in domestication, 57
coffee berry borer, 17
cognisphere, 85
Collins, Harry, 34, 68
communication at distance, 73–77
computers, in non-Western environment, 76–77
connectome, 79
controversy
    in ANT and BSE, 34–35
    and zoonotic diseases, 35, 36–37
Core Jr., 61
CRISPR (clustered regularly interspaced short palindromic repeats), 85–86
cross-species sociality, 44–45

cultural relativism, 46–47
culture
    and animals, 12, 55
    in anthropology, 52–53, 54–55
    and ethnocentrism, 3–4
    and food, 60–61
    vs. nature, 39, 45–48
cyborg anthropology, 77, 78
A Cyborg Manifesto (Haraway), 77
cyborgs
    animals as, 83–84
    and babies, 80
    and cities, 81–83
    and information, 84–85
    in military, 1, 87–88
    origins, 77–78
    and transhumanism, 77–79

D
Davies, Tony, 5
de Castro, Eduardo Viveiros, 47
dengue fever control, 33
dependence and dependency, 72
Descola, Philippe, 47–48, 54
DiNovelli-Lang, Danielle, 54
disability, and implants, 78–80
disciplines, and boundaries, 65
diseases, across species. *See* zoonotic diseases
dogs, 57, 58–59, 83, 96–97
domestication, 57–58
Dunn, Rob, 2, 19, 22

E
eating animals, 59–62
Ecuador, and IVF, 81
emotions, 21, 60–61
Enlightenment era, 4, 5, 52, 73
entanglements, 7–8, 9–10, 23, 95–96
Ento team, 61–62
Escobar, Arturo, 76, 91
ethics, in transhumanism, 89–90
ethnocentrism, 3–4, 6
ethnography, 43, 51, 56
Eurocentric view. *See* Western worldviews
evolution, and non-humans, 20

F
Facebook use, 75
Fanon, Frantz, 6

Farmer, Paul, 33
fecal matter, as drug, 22
fecal transplants, 21–22
Federal Drug Administration (FDA), 22
Fedigan, Linda, 28–29
feminism, 28–29, 78
Fidler, David P., 32
fire as tool, 69
fitness trackers, 5
Florentine Board of Health, 31
Foner, Nancy, 76
food
  animals as, 59–62
  BSE-tainted beef in, 38
  cooking of, 69
  and culture, 60–61
  and emotions, 60–61
  and microbes, 18–20, 21
  sustainability, 61–62
food safety and regulation, in BSE crisis, 38–39
food trade, in BSE crisis, 40–41
Fore people, 36–37
forests, in multinaturalism, 49–50
Franklin, Sarah, 81
future
  for humanity, 95–96
  in posthumanism, 3
  and transhumanism, 3, 67–68, 87–89

G

Gajdusek, Carleton, 36
Garreau, Joel, 67–68, 87, 88
Geertz, Clifford, 52, 54–55
gene editing, 85–86
genetics, and domestication, 58
germ-free world, 18–19
germline modification, 86
Gibson, James, 72
Ginsburg, Faye, 79
Glasse, Robert, 36
Goody, Jack, 70
grains, domestication, 57–58

H

Hakken, David, 71, 77
Hall, Stuart, 74
Halverson, John, 70
Hansen, Paul, 59
Haraway, Donna, 20, 28, 49, 58–59, 77–78

Hayles, Katherine, 84–85
Hinchliffe, Steve, 11–12
Hodder, Ian, 70–71, 72–73
holism
  in anthropology, 65
  and boundaries, 7–9, 11
  and entanglements, 7–8, 9–10, 23, 95–96
  expansion of, 7, 8–9
  and microbiome, 33
  and non-humans, 7, 8
  as principle, 7
  without boundaries, 9, 13, 65, 97–98
homes, tools and technology in, 66–67, 81–83
hominids, and posthumanism, 3
Homo sapiens, 69
Honda, Kenya, 18
Hong Kong, entanglements in SARS, 8
"host behavior manipulation," 21
human–animal relations, 5, 55–57
human exceptionalism, 44
humanism
  agency in, 23
  in Enlightenment, 4, 5, 52
  issues and biases in, 4, 5–6, 52
  and microbiome, 18
  non-humans and dualism, 4–5, 6, 23–24, 29–30, 44–46
  perfectibility of humans, 5
  vs. post-structuralism, 52
  tools and technology in, 5
  and transhumanism, 86–87
humanity and humans
  animals and dualism in, 4–5, 43–46, 47–48, 49
  in cosmologies, 47–48
  future of, 95–96
  role of non-humans, 2–3, 96–97
Human Microbiome Project (HMP), 20
human–non-human hybrids, in ANT, 24
hunting, 60
Huxley, Julian, 89
hybrids, in ANT, 24

I

"imagined communities," 74
immunity to disease, 31
imperialism, 6
implants, and disability, 78–79

indigenous peoples
  animals and hunting, 60
  anthropology of, 9, 23, 51
  multinaturalism, 47
  postcolonialism, 54
infectious diseases, across species. *See* zoonotic diseases
information, and cyborgs, 84–85
Ingold, Tim, 8, 30, 45–46, 49, 60
inoculation, as hybrid, 24
insect eating, 61–62
institutions, and holism, 7
intentionality, 24, 26
The Interpretation of Cultures (Geertz), 52
invasive species, and borders, 10
invitro fertilization (IVF), 80–81

J
jaguars, 49–50
Jamison, Douglas W., 20
Japan, dogs in, 59
Joy, Melanie, 60

K
Kim, Heup Young, 90–91
kinship, and reproductive technologies, 80–81
Kohn, Eduardo, 25, 26–27, 49–50
Kuhn, Thomas, 35
kuru disease, 36–37
Kurzweil, Ray, 67, 89

L
language, as tool, 69–70
Lanier, Jaron, 88
Larkin, Brian, 76
Lash, Scott, 53
Latour, Bruno, 22–23, 29, 35
learning disabilities, 79
Lévi-Strauss, Claude, 69
Lien, Marianne Elisabeth, 38
Lindenbaum, Shirley, 36
linkages, 23, 24
literacy, 70
Littman, Dan R., 18
Liu, Cixin, 91–92
local societies and groups, and holism, 7–8

M
mad cow disease. *See* BSE and crisis
material realm, and agency, 26, 30–31, 41

matsutake mushrooms, 50
meat eating, 59–62
media
  and change, 77
  and communication at distance, 73–77
  digital distribution, 74–75
  impact on humans, 74–76
  in non-Western environment, 76–77
miasma theory, 27
microbe-free world, 18–19
microbes
  fecal transplants, 21–22
  and food, 18–20, 21
  incorporation in body, 20, 22
microbiome
  and behavior, 21
  benefits of, 19, 22
  and borders, 17, 31–33
  capacities and agency, 27
  in digestion, 18
  and holism, 33
  and humanism, 18
  importance and role, 1, 18
  incorporation in body, 20
  inner and outer view, 11–12
  manipulation of, 17, 19
  and modern mobility, 32
  and obesity, 19–20
  and posthumanism, 22–23
  research and investment in, 19–20
  and sanitization, 18–19, 22
migrants, and communication, 75, 76
mind, in animals and humans, 48
Mintz, Sidney, 98
mitochrondria, 20
Moore, Michael, 55
more-than-human elements, 2–3, 9, 68–70
movies, and technology, 67–68
multiculturalism, 47, 49–50
multinaturalism, 12, 47, 49–50
multispecies ethnography, 56
multivocality, 53

N
Nading, Alex, 33
naturalism, 47–48
natural sciences, 49
nature
  vs. rationality and culture, 45–48, 49
  and technology, 81

naturecultures, 49, 59
Nauert, Charles, 4
Neolithic Revolution, 57
neuroscience, 79–80
new media, 74–76
new reproductive technologies, 80–81
non-humans
    actants and agency, 23–27, 29–30, 71
    in anthropology, 50, 52–53, 54–55, 56
    and borders, 10–11
    and entanglements, 7–8, 95–96
    and evolution, 20
    and holism, 7, 8
    humanism and dualism in, 4–5, 6, 23–24, 29–30, 44–46
    impact of, 11
    linkages in posthumanism, 23
    in multinaturalism, 50
    in religion, 45
    role in humans and humanity, 2–3, 96–97
    See also animals
non-living objects, and agency, 25
non-Western world and views, 76–77, 90–91

O

obesity epidemic, 19–20
objects, impact as technology, 71–72
oil sands, and actants, 24–25
ontologies, 48, 54
open wholes, 50

P

pandemics, and biosecurity, 32–33
Paralympics, and technology, 78–79
past, in posthumanism, 2–3, 68–70, 91
Pasteur, Louis, 18
path dependency, 72–73, 88
perfectibility of humans, 5
personhood, 46, 47, 52
Peters, Ted, 90
pets, as family, 58–59
Pickering, Andrew, 26, 29, 30–31, 35, 44
Pinch, Trevor, 34, 68
Pistorius, Oscar, 78
plague, and borders, 31–32
plants, domestication, 57–58
polyvocality, 53
post-anthropocentric path, 4–6, 12, 29, 48, 62, 96, 98

postcolonialism, 53–54
posthumanism
    on animals-humans dualism, 47–48
    and anthropology, 6, 9, 52
    ethics, 89
    future in, 3
    importance, 10
    and linkages with non-humans, 23
    and microbiome, 22–23
    past in, 2–3, 68–70, 91
    and postcolonialism, 54
    rejection of anthropocentrism, 4–6, 12, 48, 62, 96, 98
    role of, 6, 44
    and science, 27–28
    technology in, 2, 68–69
    vs. transhumanism, 3, 4, 91–92
posthumans, study of, 4
postmodernism, 50, 51–53, 54–55
post-structuralism, 52
President's Council on Bioethics, 87
primatology, as feminist science, 28
prions (proteinaceous infectious particles), 35–36, 37, 38
probiotics, 21
prosthetic extensions, 68–70, 78–79
Prusiner, Stanley, 35–36, 37
Purdey, Mark, 37

Q

Quammen, David, 33–34
quarantines, 31–32
QWERTY keyboard, 72–73

R

Rapp, Rayna, 79–80
rationality, in Enlightenment, 4, 5, 52
religion, 13, 45, 89, 90–91
Reyniers, James, 18
Ridley Inc., 39
Roberts, Elizabeth, 80–81
Robinson, Charlotte, et al., 83
Roseberry, William, 55

S

Sabbath, and technology, 83
Sahlins, Marshall, 60–61
salvage anthropology, 51
SARS (severe acute respiratory sydrome), entanglements in, 8
scallop industry, 25, 49

science
    and agency, 29–31, 49
    attitudes towards, 68
    bias in, 28
    "black box" idea, 35
    and controversy, 35, 37
    and feminism, 28–29
    and posthumanism, 27–28
    and sociology, 29–30
    See also technology
Science Wars, 53
scrapie disease, 35–36, 37
Seventure, 20
sexist bias in science, 28
Singularity, 67
small-scale societies, in anthropology, 9, 23, 51
smart homes, 66–67, 81–83
sociality, across species, 44–45
social media, 74, 75
sociology, 23, 29–30
soldiers, future of, 1, 87–88
Solomon, Olga, 44–45, 52
soul, in animals and humans, 48
spirituality, and transhumanism, 13
spoken language, as tool, 69–70
Stephenson, Neal, 77–78
Strengers, Yolande, 83
surgical implants, 78

T

technological determinism, 70–73
technology
    adaptation to, 82
    and anthropology, 1–2
    attitudes towards, 68
    body–technology mergers (*See* transhumanism)
    and change, 70–73, 76
    in cognisphere, 85
    cyborgs (*See* cyborgs)
    fear of, 87–88
    and friendship, 75
    for the future, 96–97
    and human brain, 1–2
    in humanism, 5
    impact of "things," 71–72
    and nature, 81
    in posthumanism, 2, 68–69
    transformation in, 66–68
    use by humans, 96–97

    utopia and reality, 82–83, 89–90
    and work, 88
texts, in anthropology, 52
"things," impact as technology, 71–72
3D printing, 9
tools
    in homes, 66–67, 81–83
    and humanism, 5
    as prosthetic extensions, 68–70
tourism, entanglements in SARS, 8
Toxoplasma gondii, 21
toxoplasmosis, 21
transhumanism
    and cyborgs, 77–79
    ethics in, 89–90
    examples and trends in, 2
    as future, 3, 67–68, 87–89
    and humanism, 86–87
    and new media, 75
    non-Western versions, 90–91
    opposition to and fear of, 87–88
    perfectibility of humans, 5
    vs. posthumanism, 3, 4, 91–92
    proponents of, 87
    and religion, 13, 89, 90–91
    as social advantage, 90
    utopia and reality, 89–90
transmissible spongiform encephalopathies (TSEs), 35, 37
Tsing, Anna, 9–10, 50, 57
Tufekci, Zeynep, 75
Turnbaugh, Peter, et al., 20

U

Udell, Monique, and colleagues, 58
United States, 22, 51, 59, 60
US Department of Defense, 1, 87
US National Institutes for Health, 20

V

variant Creutzfeldt-Jakob disease (vCJD), 11, 38
Venturini, Tommaso, 34
vitamin K, and babies, 19

W

Watt, Ian, 70
Weinstone, Ann, 91
Western worldviews
    on animals, 12, 45–46
    biases of, 4, 5–6

body and soul in, 47, 48
multiculturalism, 47
Wolfe, Cary, 4, 45, 59–60, 68–69
women, sexist bias in science, 28
Woodward, Kath, 11–12
work, and technology, 88
World Health Organization (WHO), pandemics and biosecurity, 32
Wrangham, Richard, 2, 69
written language, as tool, 69–71

Y
Young, Simon, 87

Z
Zimmer, Carl, 20
zoonotic diseases (zoonoses)
and biosecurity, 33
and borders, 17, 31
and controversy, 35, 36–37
and epidemics, 11
prevalence, 33–34
See also BSE and crisis; specific diseases